STAGE-COACH
AND MAIL IN
DAYS OF YORE

A PICTURESQUE HISTORY
OF THE COACHING AGE

VOL. II

By CHARLES G. HARPER

*Illustrated from Old-Time Prints
and Pictures*

LONDON :

CHAPMAN & HALL, LIMITED

1903

PRINTED BY
HAZELL, WATSON AND VINEY, LD,
LONDON AND AYLESBURY

CONTENTS

LIST OF ILLUSTRATIONS

SEPARATE PLATES

ILLUSTRATIONS IN TEXT

STAGE
COACH
AND
MAIL

IN
DAYS
OF
YORE

CHAPTER I

THE LATER MAILS

THE Bristol Mail opened the mail-coach era by
going at eight miles an hour, but that was an
altogether exceptional speed, and the average
mail-coach journeys were not performed at a
rate of more than seven miles an hour until long
after the nineteenth century had dawned. In
1812, when Colonel Hawker travelled to Glasgow,
it took the mail 57 hours' continuous unrelaxing
effort to cover the 404 miles—three nights and
two days' discomfort. By 1836 the distance
had been reduced by eight miles, and the time
to 42 hours. By 1838 it was 41 hours 17 minutes.
Nowadays it can be done by quickest train in
exactly eight hours; the railway mileage 401½
miles. In 1812 it cost an inside passenger all
the way to Glasgow, for fare alone, exclusive of
tips to coachmen and guards, and the necessary
expenditure for food and drink all those weary

hours, no less than £10 8s.; about 6⅛d. a mile. To-day, £2 18s. franks you through, first-class; or 33s. third—itself infinitely more luxurious than even the consecrated inside of a mail-coach.

The mails starting from London were perfection in coaches, harness and horses; but as the distance from the Metropolis increased so did the mails become more and more shabby. Hawker, travelling north, found them slow and slovenly, the harness generally second-hand, one horse in plated, another in brass harness; and when they *did* have new (which, he tells us, was very seldom) it was put on like a labourer's leather breeches, and worn till it rotted, without ever being cleaned.

Of course, very few people ever did, or could have had the endurance to, travel all that distance straight away, and so travel was further complicated, delayed, and rendered more costly by the halts necessary to recruit jaded nature.

Hawker evidently did it in four stages: to Ferrybridge, 179 miles, where he rested the first night and picked up the next mail the following; thence the 65 miles onward to Greta Bridge; on again, 59 miles, to Carlisle; and thence, finally, to Glasgow in another 101 miles. In his diary he gives "a table to show for how much a gentleman and his servant (the former inside, with 14 lb. of luggage; the latter outside, with 7 lb.) may go from London to Glasgow."

			£ s. d	£ s d
Self.				
Inside, to Ferrybridge .	. .	.	4 16 0	
„ „ Greta Bridge	. .		1 12 6	
„ „ Carlisle	. .		1 9 6	
„ „ Glasgow	. .	.	2 10 0	
				10 8 0
Servant.				
Outside, to Ferrybridge	. .	.	2 10 0	
„ „ Greta Bridge	.	.	1 2 0	
„ „ Carlisle	. .	.	1 0 0	
„ „ Glasgow	. .	.	1 13 0	
				6 5 0
Tips.				
Inside, 6 long-stage coachmen @ 2s.	.		0 12 0	
„ 12 short-stage coachmen @ 1s	.		0 12 0	
„ 7 guards @ 2s each	.		0 14 0	
Outside, for man, @ half price above			0 19 0	
				2 17 0
Total	.	.	.	£19 10 0

Such were the costs and charges of a gentleman travelling to pay a country visit in 1812,
exclusive of hotel bills for self and servant on
the way

The great factor in the acceleration of the
mails was the improvement in the roads, a
work carried out by the Turnpike Trusts in
fear of the Post Office, whose surveyors had the
power, under ancient Acts, of indicting roads in
bad condition. Great bitterness was stirred up
over this matter. The growing commercial and
industrial towns—Glasgow prominent among
them—naturally desired direct mail-services, and
the Post Office, using their needs as means
for obtaining, not only roads kept in good condition, but sometimes entirely new roads and
short cuts, declined to start such services until
such routes were provided. It was not within
the power of the Department to compel new

roads, but only to see that the old ones were maintained; but in the case of Glasgow, to whose merchants a direct service meant much, the Corporation, the Chamber of Commerce, and individual persons contributed large sums for the improvement of the existing road between that city and Carlisle, and a Turnpike Trust was formed for one especial section, where the road was entirely reconstructed. These districts were wholly outside Glasgow's sphere of responsibilities, but all this money was expended for the purpose of obtaining a direct mail through Carlisle, instead of the old indirect one through Edinburgh; and when obtained, of retaining it in face of the continued threats of the Post Office to take it off unless the road was still further improved. It certainly does not seem to have been a remarkably good road, even after these improvements, for Colonel Hawker, travelling it in 1812, describes it as being mended with large soft quarry-stones, at first like brickbats and afterwards like sand.

But the subscribers who had expended so much were naturally indignant. They pointed out that the district was a wild and difficult one and the Trust poor, in consequence of the sparse traffic. The stage-coaches, they said, had in some instances been withdrawn because they could not hold their own against the competition of the mail, and the Trust lost the tolls in consequence; while the mail, going toll-free and wearing the road down, contributed nothing to the upkeep. They urged that the mail should at least pay toll, and in this they were supported by every other Turnpike Trust.

The exemption of mail-coaches from payment of tolls, a relief provided for by the Act of 25th George III., was really a continuation of the old policy by which the postboys of an earlier age, riding horseback and carrying the mailbags athwart the saddle, had always passed toll-free. Even the light mail-cart partook of this advantage, to which there could then have been no real objection. It had been no great matter, one way or the other, with the Turnpike Trusts, for the posts were then infrequent and the revenue to be obtained quite a negligeable quantity; but the appearance of mail-coaches in considerable numbers, running constantly and carrying passengers, and yet contributing nothing towards the upkeep of the roads, soon became a very real grievance to those Trusts situated on the route of the mails, but in outlying parts of the kingdom, little travelled, and where towns were lacking and villages poor, few, and far between. Little wonder, then, that the various Turnpike Trusts in 1810 approached Parliament for a redress of these disabilities. They pointed out that not only was there a greater wear and tear of the roads now the mail-coaches were running, but that travellers, relying on the fancied security of the mails, had deserted the stages, which in many cases had been wholly run off the road. Pennant, writing in 1792, tells how two

stages plying through the county of Flint, and
yielding £10 in tolls yearly, had been unable
to compete with the mail, and were thus
withdrawn, to the consequent loss of the Trust
concerned. It was calculated, so early as 1791,
by one amateur actuary, that the total annual
loss through mail exemptions was £90,000; but
another put it at only £50,000 in 1810.

The case of the remote country trusts was
certainly a hard one   Like all turnpikes, they
were worked under Acts of Parliament, which
prescribed the amounts of tolls to be levied, and
they were, further, authorised to raise money for
the improvement of the roads on the security of
the income arising from these taxes upon locomo-
tion. The security of money sunk in these quasi-
Government enterprises had always been considered
so good that Turnpike Trust bonds and mortgages
were a very favourite form of investment; but
when Parliament turned a deaf ear to the bitter
cry of the remote Trusts, the position of those
interested in the securities began to be recon-
sidered. The woes of these undertakings were
further added to by the action of the Post Office,
which, zealous for its new mail-services, sent out
emissaries to report upon the condition of the
roads. The reports of these officials bore severely
against the very Trusts most hardly hit by the
mail-exemption, and the roads under their control
were frequently indicted for being out of repair,
with the result that heavy fines were inflicted.
It had been suggested that as the Post Office on

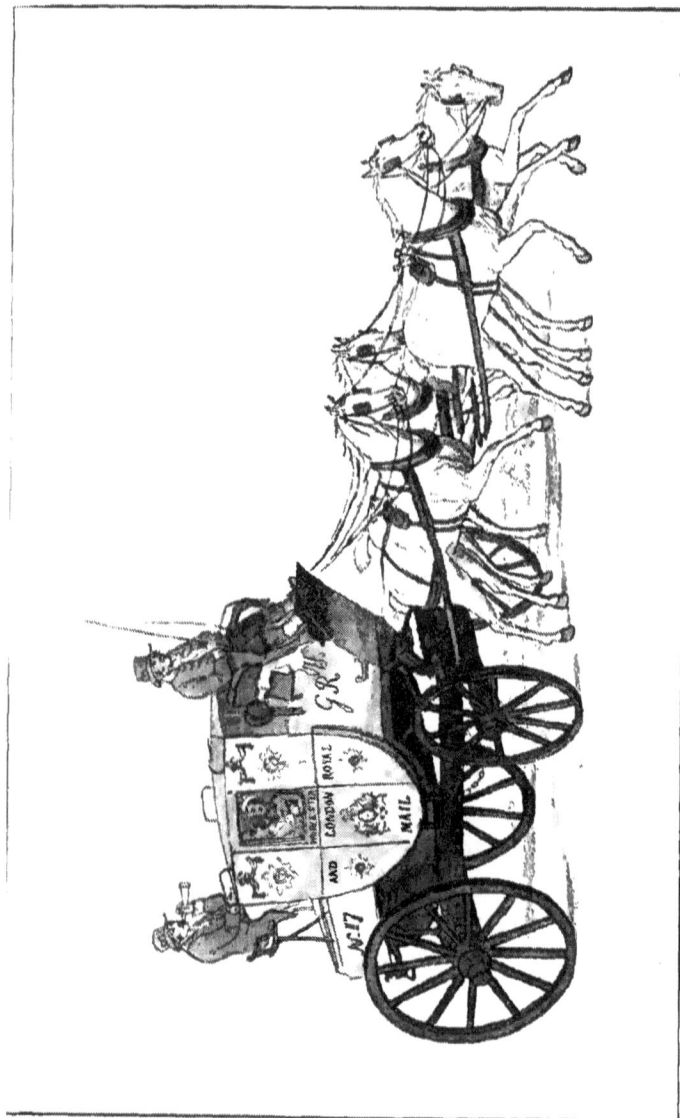

THE WORCESTER MAIL, 1805.

After J. A. Atkinson.

one hand required better roads, and on the other deprived the rural Trusts of a great part of their income, the Government should at least pay off the debts of the various turnpikes. But nothing was done; the mails continued to go free, and in the end the iniquity was perpetrated of suffering the local Turnpike Acts to lapse and the roads to be dispiked before the Trusts had paid off their loans. The greater number of Trust "securities" therefore became worthless, and the investors in them ruined.

Mail-coaches continued to go toll-free to the very last in England, although from 1798 they had paid toll in Ireland. In Scotland, too, the Trusts were treated with tardy justice, and in 1813 an Act was passed repealing the exemption in that kingdom. But what the Post Office relinquished with one hand it took back with the other, clapping on a halfpenny additional postage for each Scotch letter. It was like the children's game of "tit-for-tat." But it did not end here. The Trusts raised their tolls against the mail-coaches, and smiled superior. It was then the Department's call, and it responded by immediately taking off a number of the mails. That ended the game in favour of St. Martin's-le-Grand.

Although Parliament never repealed the exemption for the whole of the United Kingdom, it caused an estimate to be prepared of the annual cost of paying tolls, should it ever be in a mind to grant the Trusts that relief. It thus appeared, from the return made in 1812, that the cost for Scotland would have been £11,229 16s. 8d ; for England, £33,536 2s. 3d.; and for Wales, £5224 3s. 10d. : total, £49,990 2s. 9d per annum.

The mails, travelling as they did throughout the night, were subject to many dangers. They were brilliantly lighted, generally with four, and often with five, lamps, and cast a very dazzling illumination upon the highway. It is true that no certainty exists as to the number of lamps mail-coaches carried, and that old prints often show only two ; so that the practice in this important matter probably varied on different routes and at various times. But the crack mails at the last certainly carried five lamps—one on either side of the fore upper quarter, one on either side of the fore boot, and another under the footboard, casting a light upon the horses' backs and harness. These radiant swiftnesses, hurtling along the roads at a pace considerably over ten miles an hour, were highly dangerous to other users of the roads, who were half-blinded by the glare, and, alarmed by the heart-shaking thunder of their approach and fearful of being run down, generally drove into the ditches as the least of two evils. The mails were then, as electric tramcars and high-powered motor-cars are now, the tyrants of the road.

But they were not only dangerous to others. Circumstances that ought never to have been permitted sometimes rendered them perilous to all they carried. The possibilities of that time in wrong-doing are shown in the practice of Sir Watkin Williams Wynn (who assuredly was not the only one) being allowed to send his refractory carriage-horses to the mails, to be steadied. On such occasions the passengers from Oswestry found themselves in for a wild start and a rough stage, and Sir Watkin had the steam taken out of his high-mettled horses at an imminent risk to the lives and limbs of the lieges.

From 1825, when the era of the fast day-coaches began, the mails gradually lost the proud pre-eminence they had kept for more than forty years. Even though they had been accelerated from time to time as roads improved, they went no quicker than the new-comers, and very often not so quick, from point to point. They suffered the disabilities of travelling by night, when careful coachmen dared not let their horses out to their best speed, and of being subject to the delays of Post Office business; and so, although they might, and did, make wonderful speed between stages, the showing on the whole journey could not compare with the times of the fast day-coaches, which halted only for changing horses and for meals, and, enjoying the perfection of quick-changing, often got away in fifty seconds' from every halt. Going at more seasonable hours, the day-coaches now began to seriously compete with the mails, whose old-time supporters, although still sensible of the dignity of travelling by mail, were equally alive to the comfort and convenience of going by daylight. Modern writers, enlarging upon the times of our ancestors, lay great stress upon the endurance our hearty grandfathers " cheerfully " displayed, and show us great, bluff, burly, red-cheeked men, who enjoyed this long night-travelling. But that is an absurdity. They did *not* enjoy it; they were not all bluff and burly, and that they welcomed the change that gave them swift travelling by day instead of night is obvious from the instant success of the fast day-coaches, and from the later business-history of the mails. Mail-contractors, who in the prosperous days of no competition were screwed down by the Post Office to incredible mileage figures, began to grumble; but for long they grumbled in vain. Even in 1834 the Post Office continued to pay only 2*d*. a mile on 42 mails, 1½*d*. a mile on 34, and only one received as much as 1*d*. The Liverpool and Manchester carried the mailbags for nothing, and three actually paid the Post Office for the privilege. At this time the old rule forbidding more than three outside passengers on the mails was relaxed. This indulgence began in Scotland, where the contractors, in consideration of the sparseness of the population and the scarcity of chance passengers on the way, were allowed a fourth outside passenger; and eventually many of the mails, like the stages, carried from eight to twelve outsides. This, however, did not suffice, for those passengers did not often present themselves, and at last the contractors really did not care to obtain the Post Office business, finding it pay better to devote their

THE MAIL.

After J. L. Agasse, 1824.

attention to fast day-coaches on their own
account.

Thus the Post Office found itself in a novel
and unwonted position. Coach-proprietors and
contractors, instead of anxiously endeavouring to
obtain the mail-contracts, held aloof, and the Post
Office surveyors, when renewals were necessary,
found *they* had to make the advances and do the
courting. Then the tables were turned with a
vengeance! For Benjamin Horne's "Foreign
Mail," carrying what were called the "black
bags" (*i.e.* black tarpaulin to protect the mail
from sea-water) between London and Dover,
1*s*. 3¾*d*. per double mile was paid; 11⅓*d*. for the
Carmarthen and Pembroke; and 8*d*., and then
9*d*., for the Norwich Mail, by Newmarket,
strongly opposed as it was by the Norwich "Telegraph," and therefore loading badly on that lonely
road. For the Chester, originally contracted for
at 1*s*. a mile, then down to 3*d*., and in 1826 up
to 4*d*, 6*d*. was paid, on account of passengers
going by the direct Holyhead Mail, and the
Holyhead itself was raised to the same figure
when fast day-stages had begun to run from
Shrewsbury.

A Committee of the House of Commons had
sat upon this question before these prices were
given, and much evidence was taken; but these
revised tariffs did by no means end the matter.
Substantial contractors would in many instances
have nothing to do with the Post Office, and the
Department could not run the risk of employing

irresponsible men who could not be held to
their undertakings. In some few instances
ordinary night-stages were given the business,
and it was seriously proposed to employ the
guards of existing stage-coaches to take charge
of the bags, but this was never carried out. In
the midst of all these worries, when it seemed as
though the despatch of the mails must needs, in
the altered conditions of the time, be eventually
changed from night to day, railways came to
relieve official anxieties, which existed not only
on account of the increasing cost, but also on the
score of the continually growing bulk of mail-
matter, piled up to mountainous heights on the
roof, instead of, as originally, being easily stowed
away in the depths of the hind boot It was
considered a great advantage of the mail-coaches
built by Waude in these last days that they were
not only built with a low centre of gravity, but
that, with a dropped hind axle, they made a
deeper and more capacious boot possible, in which
were stowed the more valuable portions of the
mail. Had railways not at the very cynthia of
the moment come to supply a "felt want," certainly the mails must on many roads have been
carried by mail-vans devoted exclusively to the
service But in 1830 the Liverpool and Manchester Railway carried mailbags, and in anticipation of the opening throughout of the London
and Birmingham, the first long route, in September
1838, an Act of Parliament was passed on
August 14th in that year, authorising the

conveyance of mails by railways. We must not, however, suppose that such instant advantage was always taken of new methods. That would not be according to the traditions of the Post Office. Accordingly, we find that, although what is now the London and South-Western Railway was opened between Nine Elms and Portsmouth in May 1840, it was not until 1842 that the Portsmouth Mail went by rail. For two years it continued to perform the 73 miles 3 furlongs in 9 hours 10 minutes, when it might have gone by train in 6 hours 10 minutes less.

With these changes, London lost an annual spectacle of considerable interest. From 1791 the procession of the mail-coaches on the King's birthday had been the grand show occasion of the Post Office year. No efforts and no expense were spared by the loyal contractors (loyal in spite of the ofttimes arbitrary dealings of the Post Office with them) to grace the day; and Vidler and Parratt, who for many years had the monopoly of supplying the coaches, equalled them in the zeal displayed. The coaches were drawn up at twelve, noon, to the whole number of twenty-seven, at the factory on Millbank, beautiful in new paint and new gilding; the Bristol Mail, as the senior, leading, the others in the like order of their establishment. On this occasion the Post Office provided each guard with a new gold-laced hat and scarlet coat, and the mail-contractors who horsed the coaches, not to be outdone, found scarlet coats for their coachmen, in addition to providing new

harness. The coachmen and guards, unwilling
to be beaten in this loyal competition, provided
themselves with huge nosegays, as big as cauli-
flowers. When, as in the reign of William IV,
the King's birthday fell in a pleasant time
of the year, the procession of the mails was
a beautiful and popular sight, attracting not
only the general public, but the very numerous
sporting folks, who welcomed the opportunity of
seeing at their best, and all together, the one
hundred and two noble horses that drew the
mails from the Metropolis to all parts of the
kingdom. Everything, indeed, was very special
to the occasion. Each coach was provided with
a gorgeous hammer-cloth, a species of upholstery
certainly not in use on ordinary journeys. No one
was allowed on the roof, but the coachman and
guard had the privilege of two tickets each for
friends for the inside. Great, as may be supposed,
was the competition for these. For the contractors themselves there was the cold collation
provided by Vidler and Parratt at Millbank, at
three o'clock, when the procession was over.

The route varied somewhat with the circumstances of the time, always including the residence
of the Postmaster-General for the time being.
Punctually at noon it started off, headed by a
horseman, and with another horseman between
each coach. Nearing St. James's Palace, it was
generally reduced to a snail's pace, for the crowd
always assembled densely there, on the chance
of seeing the King, and the authorities of that

THE BRISTOL MAIL AT HYDE PARK CORNER, 1838.

After J. Doyle.

period were not clever at clearing a route.
Imagine now the front of Carlton House Palace,
or St. James's, and the Londoners of that age
assembled in their thousands. The procession
with difficulty approaches, and halts. Two barrels
of porter—Barclay & Perkins' best—are in
position in front of the Royal residence, and to
each coachman and guard is handed a capacious
pewter pot—it is a sight to make a Good Templar
weep The King and Queen and the Royal
family now appear at an open window, the King
removing his hat and bowing, to a storm of
applause—as though he had done something
really clever or wonderful. Now the coachman
of the Bristol Mail uncovers, and holding high
the shining pewter, exclaims "We drink to
the health of His Gracious Majesty : God bless
him!" and suiting the action to the words, dips
his nose into the pot, which in an incredibly
short time is completely inverted and emptied.
Fifty-three other voices simultaneously repeat
the same words, and fifty-three pint pots are
in like manner drained in the twinkling of an
eye. The King and his family now retire, and
the procession prepares to move on; but the
mob, moved by loyalty and the sight of the
beer-barrels, grows clamorous : "King, King!
Queen, Queen!" cry a thousand voices; while
a thousand more yell, "Beer, beer!" When at
length the King does return, to bow once more,
he gazes upon two thousand people struggling
for two half-empty barrels, which in the scuffle

have upset, and speedily become empty. Meanwhile the coaches have moved off, to complete their tour to the General Post Office, and thence back to Millbank.

These processions, from some cause or another not now easily to be discovered, were omitted in 1829 and 1830. May 17th, 1838, when twenty-five mails paraded, was the last occasion; for already the railway was threatening the road, and when Queen Victoria's birthday recurred the ranks of the mails were sadly broken.

This memorable year, 1837, then, was the last unbroken year of the mail-coaches starting from London Since September 23rd, 1829, when the old General Post Office in Lombard Street was deserted for the great building in St. Martin's-le-Grand, they had come and gone. The first ever to enter its gates, as the result of keen competition, had been the up Holyhead Mail of that date; the last was the Dover Mail, in 1844.

The mail-coaches loaded up about half-past seven at their respective inns, and then assembled at the Post Office Yard to receive the bags. All, that is to say, except seven West of England mails—the Bath, Bristol, Devonport, Exeter, Gloucester, Southampton and Stroud—whose coaches started from Piccadilly, the bags being conveyed to them at that point by mail-cart. There were thus twenty-one coaches starting nightly from the General Post Office precisely at 8 o'clock. Here is a list of the mails setting out every night throughout the year :—

Mails.	Miles.	Inn whence starting	Time.		Average speed per hour, stops included.	
			H	M	M	F
Bristol	122	Swan with Two Necks	11	45	10	3
Devonport ("Quicksilver")	216	Spread Eagle, Gracechurch Street	21	14	10	1¾
Birmingham	119	King's Arms, Holborn Bridge	11	56	9	7¾
Bath	109	Swan with Two Necks	11	0	9	7¼
Manchester	187	" "	19	0	9	6⅔
Halifax	196	" "	20	5	9	6
Liverpool	203	" "	20	50	9	6
Holyhead	261	" "	26	55	9	5⅔
Norwich, by Ipswich	113	" "	11	38	9	5⅔
Exeter	173	" "	18	12	9	4
Hull (New Holland Ferry)	172	Spread Eagle, Gracechurch Street	18	12	9	4
Leeds	197	Bull and Mouth	20	52	9	3½
Glasgow	396	" "	42	0	9	3⅗
Southampton	80	Swan with Two Necks	8	30	9	3⅓
Edinburgh	399	Bull and Mouth	42	23	9	3⅓
Chester	190	Golden Cross	20	16	9	3
Gloucester and Carmarthen	224	" "	24	0	9	2⅔
Worcester	115	Bull and Mouth	12	20	9	2¼
Yarmouth	124	White Horse, Fetter Lane	13	30	9	1½
Louth	148	Bell and Crown, Holborn	15	56	9	0
Norwich, by Newmarket	118	Belle Sauvage	13	5	9	0
Stroud	105	Swan with Two Necks	11	47	9	0
Wells	133	Bell and Crown	14	43	9	0
Falmouth	271	Bull and Mouth	31	55	8	4
Dover	73	Golden Cross	8	57	8	1¼
Hastings	67	Bolt-in-Tun, Fleet Street	8	15	8	0
Portsmouth	73	White Horse	9	10	7	7 1/7
Brighton	55	Blossoms Inn	7	20	7	4

With the exception of the Brighton, Portsmouth, Dover and Hastings, they were all splendidly-appointed four-horse coaches ; but those four places being only at short distances, speed was unnecessary, and they were only provided with pair-horse mails. Had a speed similar to that maintained on most other mails been kept up, letters and passengers would have reached the coast in the middle of the night.

The so-called " Yarmouth Mail " was, we are told by those who travelled on it, an ordinary stage-coach, carrying the usual four inside and twelve outside, chartered by the Post Office to carry the mail-bags ; but the old print, engraved here, does not bear out that contention.

The *arrival* of the mails in London was an early morning affair. First of all came the Leeds, at five minutes past four, followed at an interval of over an hour—5.15—by the Glasgow, and then, at 5.39, by the Edinburgh. All arrived by 7 o'clock.

There was then, as now, no Sunday delivery of letters in London, and Saturday nights were, by consequence, saturnalias for the up-mails. Although the clock might have been set with accuracy by their passing at any other time, their coming into London on Sundays was a happy-go-lucky, chance affair. The coachmen would arrange to meet on the Saturday nights at such junctions of the different routes as Andover, Hounslow, Puckeridge, and Hockliffe, and so in company have what they very descriptively termed a " roaring time."

THE YARMOUTH MAIL AT THE " COACH AND HORSES," ILFORD.   *After J. Pollard.*

In 1836 the fastest mail ran on a provincial route. This was the short 28-miles service between Liverpool and Preston, maintained at 10 miles 5 furlongs an hour. The slowest was the 19-miles Canterbury and Deal, at 6 miles an hour, including stops for changing. The average speed of all the mails was as low as 8 miles 7 furlongs an hour.

In 1838 there were 59 four-horse mails in England and Wales, 16 in Scotland, and 29 in Ireland, in addition to a total number of 70 pair-horse : some 180 mails in all. It was in this year that—the novelty of railways creating a desire for fast travelling—the Post Office yielded to the cry for speed, and, abandoning the usual conservative attitude, went too far in the other direction, overstepping the bounds of common safety. For some time the mails between Glasgow and Carlisle, and Carlisle and Edinburgh were run to clear 11 miles an hour, which meant an average pace of 13 miles an hour. These were popularly called the " calico mails," because of their lightness, The time allowed between Carlisle and Glasgow, 96 miles, was 8 hours 32 minutes, and it was a sight to see it come down Stanwix Brow on a summer evening. It met, however, with so many accidents that cautious folk always avoided it, preferring the orthodox 10 miles an hour—especially by lamplight in the rugged Cheviots. Even at that pace there had been more than enough risk, as these incidents from

Post Office records of three years earlier clearly show :—

1835

February	5	Edinburgh and Aberdeen Mail overturned
,,	9.	Devonport Mail overturned.
,.	10	Scarborough and York Mail overturned.
..	16.	Belfast and Enniskillen Mail overturned.
,,	,,	Dublin and Derry Mail overturned
.	17.	Scarborough and Hull Mail overturned
.,	,,	York and Doncaster Mail overturned
,,	20.	Thirty-five mail-horses burnt alive at Reading
,,	24	Louth Mail overturned.
,,	25	Gloucester Mail overturned

No place was better served by the Post Office than Exeter in the last years of the road, and few so well Before 1837 it had no fewer than three mails, and in that year a fourth was added All four started simultaneously from the General Post Office, and reached the Queen City of the West within a few hours of one another every day. On its own merits, Exeter did not deserve or need all these travelling and postal facilities, and it was only because it stood at the converging-point of many routes that it obtained them. Only one mail, indeed, was dedicated especially to Exeter, and that was the last-established, the " New Exeter," put on the road in 1837. The others continued to Devonport or to Falmouth, then a port, a mail-packet and naval station of great prominence, where the West Indian mails landed, and whence they where shipped To the mail-coaches making for Devonport and Falmouth, Exeter was, therefore, only an incident.

The "Old Exeter" Mail, continued on to
Falmouth, kept consistently to the main Exeter
Road, through Salisbury, Dorchester and Brid-
port. Before 1837 it had performed the journey to
Exeter in 20 hours and to Falmouth in 34¾ hours,
but was then accelerated one hour as between
London and Exeter, and although slightly de-
celerated onwards, the gain on the whole distance
was 19 minutes.

Five minutes in advance of this ran the
"Quicksilver" Devonport Mail, as far as Salisbury, where, until 1837, it branched off, going
by Shaftesbury, Sherborne and Yeovil, a route
5¾ miles shorter than the other. It was 1¾ hours
quicker than the "Old Exeter" as far as that
city. Here is the time-table of the "Quicksilver" at that period, to Exeter —

LEAVING GENERAL POST OFFICE AT 8 P M

Miles	Places	Due.
12	Hounslow	9 12 p m
19	Staines	9.56 ,,
29	Bagshot	11.0 ,,
67	Andover	2 42 a m
84	Salisbury	4 27 ,,
105	Shaftesbury	6 41 ,,
126	Yeovil	8 56 ,,
135	Crewkerne	10 12 ,,
143	Chard	11.0 ,,
156	Honiton	12 31 p m.
173	Exeter	2 14 ,,

Thus 18 hours 14 minutes were allowed for the
173 miles. In 1837 the "Quicksilver" was put
on the "upper road" by Amesbury and Ilminster,
and her pace again accelerated; this time by

1 hour 38 minutes to Exeter and 4 hours 39 minutes to Falmouth. This then became the fastest long-distance mail in the kingdom, maintaining a speed, including stops, of nearly $10\frac{1}{4}$ miles an hour between London and Devonport It should be remembered, when considering the subject of speed, that the mails had not only to change horses and stay for supper and breakfast, like the stage-coaches, but also had to call at the post offices to deliver and collect the mail-bags, and all time so expended had to be made up. The "Quicksilver" must needs have gone some stages at 12 miles an hour.

Time also had to be kept in all kinds of weather, and the guard—who was the servant of the Post Office, and not, as the coachman was, of the mail-contractors—was bound to see that time was kept, and had power, whenever it was being lost, to order out post-horses at the expense of the contractors. Six, and sometimes eight, horses were often thus attached to the mails. The route of the "Quicksilver" from 1837 was according to the following time-bill :—

LEAVING GENERAL POST OFFICE AT 8 P.M

Miles	Places	Due	Miles	Places	Due
12	Hounslow .	9 8 p m	97	Chicklade	5 15 a m
19	Staines	9 18 ,,	125	Ilchester .	7 50 ,,
29	Bagshot .	10 47 ,,	137	Ilminster	8 58 ,,
67	Andover .	2 20 a m	151	Honiton	11 0 ,,
80	Amesbury	3 39 ,,	170	Exeter	12 34 p m
90	Deptford Inn	4 34 ,,		Time 16 hours 34 minutes	

The complete official time-bill for the whole distance is appended :—

Time-Bill, London, Exeter and Devonport ("Quicksilver") Mail, 1837.

Contractors' Names.	Number of Passengers.		Stages.	Time Allowed.	Despatched from the General Post Office, the of , 1837, at 8 p.m.
	In.	Out.	M. F.	H. M.	Coach No. ⎰With timepiece sent out ⎱ safe, No. to . Arrived at the Gloucester Coffee-House at .
Chaplin .			⎰12 2 7 1 ⎱9 7	⎰ 2 47 ⎱	Hounslow. Staines. Bagshot. Arrived 10.47 p.m.
Company.			⎰9 1 10 1 8 0 ⎱3 5	⎰ 2 54 ⎱	Hartford Bridge. Basingstoke. Overton. Whitchurch. Arrived 1.41 a.m.
Broad . .			⎰6 7 ⎱13 7	0 39 1 19	Andover. Arrived 2.20 a.m. Amesbury. Arrived 3.39 a.m.
Ward . .			9 5	0 55	Deptford Inn. Arrived 4.31 a.m.
Davis . .			⎰0 5 ⎱6 5	⎰ 0 41	Wiley. Chicklade. Arrived 5.15 a.m. (Bags dropped for Hindon, 1
Whitmash			⎰6 6 7 0 13 4 4 1	⎰ 2 59 ⎱	Mere. [mile distant.) Wincanton. Ilchester. Cart Gate. Arrived 8.14 a.m.
Jeffery .			2 6 ⎱5 1	⎰ 0 44 ⎱	Water Gore, 6 miles from South Petherton. Bags dropped for that place. Ilminster. Arrived 8.58 a.m.
Soaring .			8 1	0 25 0 46	Breakfast 25 minutes. Dep. 9.23. Yarcombe, Heathfield Arms. Arrived 10.9 a.m.
Wheaton .			8 7	0 51	Honiton. Arrived 11 a.m.
Cockram .			⎰16 4 10 3 9 3	1 34 0 10 ⎰ 1 57	Exeter. Arrived 12.34 p.m. Ten minutes allowed. Chudleigh. Ashburton. Arrived 2.41 p.m.
Elliott . .			⎰13 2 6 6 4 0 1 7	⎰ 2 33 ⎱	Ivybridge. Bags dropped at Ridgway for Plympton, 3 furlongs distant. Plymouth. Arrived at the Post Office, Devonport, the of , 1837, at 5.14 p.m. by timepiece. At by clock.
			216 1	21 14	Coach No. ⎰Delivered timepiece arr. . ⎱ safe, No. to .

The time of working each stage is to be reckoned from the coach's arrival, and as any lost time is to be recovered in the course of the stage, it is the coachman's duty to be as expeditious as possible, and to report the horse-keepers if they are not always ready when the coach arrives, and active in getting it off. The guard is to give his best assistance in changing, whenever his official duties do not prevent it.

By command of the Postmaster-General.

GEORGE LOUIS, *Surveyor and Superintendent.*

The "New Exeter" Mail went at the moderate inclusive speed of 9 miles an hour, and reached Exeter, where it stopped altogether, 1 hour 38 minutes later than the "Quicksilver." The fourth of this company went a circuitous route down the Bath Road to Bath, Bridgewater, and Taunton, and did not get into Exeter until 3.57 p.m. Halting ten minutes, it went on to Devonport, and stopped there at 10 5 that night.

The tabulated form given on opposite page will clearly show how the West of England mails went in 1837.

The starting of the "Quicksilver" and the other West-country mails was a recognised London sight. That of the "Telegraph" would have been also, only it left Piccadilly at 5.30 in the morning, when no one was about besides the unhappy passengers, except the stable-helpers. Chaplin, who horsed the "Quicksilver" and other Western mails from town, did not start them from the General Post Office, but from the Gloucester Coffee-House, Piccadilly. The mail-bags were brought from St. Martin's-le-Grand in a mail-cart, and the City passengers in an omnibus. The mails set out from Piccadilly at 8.30 p m

It was at Andover that the "Quicksilver," from 1837, leaving its contemporary mails, climbed up past Abbot's Ann to Park House and the bleak Wiltshire downs, along a lonely road, and finally came, up hill, out of Amesbury to the most exposed part of Salisbury Plain, at

THE WEST OF ENGLAND MAILS, 1837.

Miles.	Places.	Old Exeter Mail, continued to Falmouth.	Devonport ("Quicksilver") Mail, continued to Falmouth.	New Exeter Mail.	Devonport Mail, by Bath and Taunton.
	General Post Office, London . . dep.	8.0 p.m.	8.0 p.m.	8.0 p.m.	8.0 p.m.
12	Hounslow. . arr.				9.12 ,,
19	Staines.			9.56 ,,	
23	Slough				
29	Maidenhead . .				10.40 ,,
58	Newbury . . .				1.53 a.m.
77	Marlborough . .				3.43 ,,
91	Devizes. . . .				5.6 ,,
109	Bath				7.0 ,,
149	Bridgewater . .				11.30 ,,
160	Taunton . . .				12.35 p.m.
180	Cullumpton . .				2.42 ,,
29	Bagshot		10.47 p.m.		
67	Andover		2.20 a.m.	2.42 a.m.	
84	Salisbury . . .	4.52 a.m.		4.27 ,,	
124½	Dorchester . . .	8.57 ,,		8.53 ,,	
126	Yeovil				
137	Bridport	10.5 ,,		11.0 ,,	
143	Chard				
80	Amesbury . . .		3.39 ,,		
125	Ilchester		7.50 ,,		
	Honiton		11.0 ,,	12.31 p.m.	
	EXETER . {arr. {dep.	2.59 p.m. 3.9 ,,	12.31 p.m. 12.44 ,,	2.12 ,,	3.57 ,, 4.7 ,,
210	Newton Abbot arr.				6.23 ,,
218	Totnes				7.25 ,,
190	Ashburton. . .		2.41 ,,		
214	Plymouth . . .		5.5 ,,		
	DEVONPORT {arr. {dep.		5.14 ,, 5.41 ,,		10.5 ,,
234	Liskeard . arr.		7.55 ,,		
246	Lostwithiel . .		9.12 ,,		
252	St. Austell . .		10.20 ,,		
266	Truro		11.55 ,,		
271	FALMOUTH . .	3.55 a.m.	1.5 a.m.		
		31 h. 55 m.	29 h. 5 m.	18 h. 12 m.	26 h. 5 m.

Stonehenge, in the early hours of the morning. The "Quicksilver" was a favourite subject with the artists of that day, who were never weary of pictorially representing it They have shown it passing Kew Bridge, and the old "Star and Garter," on the outward journey, in daylight— presumably the longest day in the year, because it did not reach that point until 9 p.m. Two of them have, separately and individually, shown us the famous attack by the lioness in 1816, and two others have pictured it on the up journey, passing Windsor Castle, and entering the City at Temple Bar; but no one has ever represented the "Quicksilver" passing beneath that gaunt and storm-beaten relic of a prehistoric age, Stonehenge. One of them, however, did a somewhat remarkable thing. The picture of the "Quicksilver" passing within sight of Windsor was executed and published in 1840, two years after the gallant old mail had been taken off that portion of the road, to be conveyed by railway. Perhaps the print was, so to speak, a post-mortem one, intended to keep the memory of the old days fresh in the recollection of travellers by the mail.

The London and Southampton Railway was opened to Woking May 23rd, 1838, and to Winchfield September 24th following, and by so much the travels of the "Quicksilver" and the other West-country coaches were shortened. For some months they all resorted to that station, and then to Basingstoke, when the line was opened so far.

THE "QUICKSILVER" DEVONPORT MAIL, ARRIVING AT TEMPLE BAR, 1834.

June 10th, 1839. This shortening of the coach route was accompanied by the following advertisement in the *Times* during October 1838, the forerunner of many others :—

"Bagshot, Surrey—19 Horses and harness. To Coach Proprietors, Mail Contractors, Post Masters, and Others.—To be Sold by Auction, by Mr. Robinson, on the premises, 'King's Arms' Inn, Bagshot, on Friday, November 2, 1838, at twelve o'clock precisely, by order of Mr. Scarborough, in consequence of the coaches going per Railway.

"About Forty superior, good-sized, strengthy, short-legged, quick-actioned, fresh horses, and six sets of four-horse harness, which have been working the Exeter 'Telegraph,' Southampton and Gosport Fast Coaches, and one stage of the Devonport Mail. The above genuine Stock merits the particular Attention of all Persons requiring known good Horses, which are for unreserved sale, entirely on Account of the Coaches being removed from the Road to the Railway."

In Thomas Sopwith's diary we find this significant passage : "On the 11th May, 1840, the coaches discontinued running between York and London, although the railways were circuitous." Thus the glories of the Great North Road began to fade, but it was not until 1842 that the Edinburgh Mail was taken off the road between London, York, and Newcastle. July 5th, 1847, witnessed the last journey of the mail on that storied road, in the departure of the coach from Newcastle-on-Tyne for Edinburgh. The next day the North British Railway was opened.

The local Derby and Manchester Mail was one of the last to go. It went off in October 1858 But away up in the far north of Scotland, where Nature at her wildest, and civilisation and population at their sparsest, placed physical and financial obstacles before the railway engineers, it was not until August 1st, 1874, that the mail-coach era closed, in the last journey of the mail-coach between Wick and Thurso. That same day the Highland Railway was opened, and in the whole length and breadth of England and Scotland mail-coaches had ceased to exist.

The mail-coaches in their prime were noble vehicles. Disdaining any display of gilt lettering or varied colour commonly to be seen on the competitive stage-coaches, they were yet remarkably striking. The lower part of the body has been variously described as chocolate, maroon, and scarlet. Maroon certainly was the colour of the later mails, and " chocolate " is obviously an error on the part of some writer whose colour-sense was not particularly exact; but we can only reconcile the " scarlet " and " maroon " by supposing that the earlier colouring was in fact the more vivid of the two. The fore and hind boots were black, together with the upper quarters of the body, and were saved from being too sombre by the Royal cipher in gold on the fore boot, the number of the mail on the hind, and, emblazoned on the

THE "QUICKSILVER" DEVONPORT MAIL, PASSING WINDSOR CASTLE.

After Charles Hunt, 1840.

upper quarters, four devices eloquent of the majesty of the united kingdoms and their knightly orders. There shone the cross of St. George, with its encircling garter and the proud motto, "*Honi soit qui mal y pense*"; the Scotch thistle, with the warning "*Nemo me impune lacessit*"; the sham-rock and an attendant star, with the *Quis separabit?* query (not yet resolved); and three Royal crowns, with the legend of the Bath, "*Tria juncta in uno.*" The Royal arms were emblazoned on the door-panels, and old prints show that occasionally the four under quarters had devices somewhat similar to those above. The name of each particular mail appeared in unobtrusive gold letters. The under-carriage and wheels were scarlet, or "Post Office red," and the harness, with the exception of the Royal cypher and the coach-bars on the blinkers, was perfectly plain.

One at least of the mail-coaches still sur-vives. This is a London and York mail, built by 'Waude, of the Old Kent Road, in 1830, and now a relic of the days of yore treasured by Messrs Holland & Holland, of Oxford Street. Since being run off the road as a mail, it has had a curiously varied history. In 1875 and the following season, when the coaching revival was in full vigour, it appeared on the Dorking Road, and so won the affections of Captain "Billy" Cooper, whose hobby that route then was, that he had an exact copy built. In the summer of 1877 it was running between Stratford-on-Avon and Leamington. In 1879 Mr. Charles A. R. Hoare, the banker, had it at Tunbridge Wells, and also ordered a copy. Since then the old mail-coach has been in retirement, emerging now and again as the "Old Times" coach, to empha-. sise the trophies of improvement and progress in the Lord Mayor's Shows of 1896, 1899 and 1901, in the wake of electric and petrol motor-cars, driven and occupied by coachmen and passengers dressed to resemble our ancestors of a hundred years ago.

The coach is substantially and in general lines as built in 1830. The wheels have been renewed, the hind boot has a door inserted at the back, and the interior has been relined; but otherwise it is the coach that ran when William IV. was king. It is a characteristic Waude coach, low-hung, and built with straight sides, instead of the bowed-out type common to the products of Vidler's factory. It wears, in consequence, a more elegant appearance than most coaches of that time; but it must be confessed that what it gained in the eyes of passers-by it must have lost in the estimation of the insides, for the interior is not a little cramped by those straight sides. The guard's seat on the "dickey"—or what in earlier times was more generally known as the "backgammon-board"—remains, but his sheepskin or tiger-skin covering, to protect his legs from the cold, is gone. The trapdoor into the hind boot can be seen. Through this the mails were thrust, and the guard sat throughout

MAIL-COACH BUILT BY WAUDE, 1830.

Now in possession of Messrs. Holland & Holland.

the journey with his feet on it. Immediately
in front of him were the spare bars, while above,
in the still-remaining case, reposed the indispensable blunderbuss. The original lamps, in their
reversible cases, remain. There were four of
them—one on either fore quarter, and one on
either side of the fore boot, while a smaller one
hung from beneath the footboard, just above the
wheelers. The guard had a small hand-lamp of
his own to aid him in sorting his small parcels.
The door-panels have apparently been repainted
since the old days, for, although they still
keep the maroon colour characteristic of the
mail-coaches, the Royal arms are gone, and in
their stead appears the script monogram, in
gold, " V R "

CHAPTER II

DOWN THE ROAD IN DAYS OF YORE

I.—A JOURNEY FROM NEWCASTLE-ON-TYNE TO LONDON IN 1772

IN 1773, the Reverend James Murray, Minister of
the High Bridge Meeting House at Newcastle,
published a little book which he was pleased to
call *The Travels of the Imagination, or, a True
Journey from Newcastle to London*, purporting to
be an account of an actual trip taken in 1772.
I do not know how his congregation received
this performance, but the inspiration of it was
very evidently drawn from Sterne's *Sentimental
Journey*, then in the heyday of its success and
singularly provocative of imitations—all of them
extraordinarily thin and poor. Sentimental
travellers, without a scintilla of the wit that
jewelled Sterne's pages, gushed and reflected in
a variety of travels, and became a public nuisance.
Surely no one then read their mawkish products,
any more than they do now.

Murray's book was, then, obviously, to any one
who now dips into it, as trite and jejune as the
rest of them, but it has now, unlike its fellows,
an interesting aspect, for the reason that he gives

details of road-travelling life which, once commonplace enough, afford to ourselves not a little entertainment. Equally entertaining, too, and full of
unconscious humour, are those would-be eloquent
rhapsodies of his which could only then have
rendered him an unmitigated bore It should
be noted here that although his picture of road-
life is in general reliable enough, we must by no
means take him at his word when he says he
journeyed all the way from Newcastle to London.
We cannot believe in a traveller making that
claim who devotes many pages to the first fifteen
miles between Newcastle and Durham, and yet
between Durham and Grantham, a distance of a
hundred and fifty miles, not only finds nothing of
interest, but fails to tell us whether he went by the
Boroughbridge or the York route, and mentions
nothing of the coach halting for the night between
the beginning of the journey at Newcastle, and
the first specified night's halt at Grantham, a
hundred and sixty-five miles away. Those were
the times when the coaches inned every night,
and not until the "Wonder" London and
Shrewsbury Coach was started, in 1825, did any
coach ever succeed in doing much more than a
hundred miles a day. So much in adverse
criticism. But while a very casual glance is
sufficient to expose his pretensions of having made
the entire journey in this manner, it is equally
evident that he knew portions of the road, and
that he was conversant with the manners and
customs that then obtained along it—as no one

then could help being. The fare between Newcastle and London, the lengthy halts on the way,
and the manner in which the passengers often
passed the long evenings at the towns where they
rested for the night—witnessing any theatrical
performance that offered—are extremely interesting, as also is the curious sidelight thrown
upon the fact that actors—technically, in the
eyes of the law, " rogues and vagabonds "—were
then actually so regarded. How poorly considered
the theatrical profession then was, is, of course,
well known ; but it is curious thus to come upon
a reference to the fact that London theatres then
had long summer vacations, in which the actors
and actresses must starve if they could not
manage to pick up a meagre livelihood by barn-
storming in the country ; as here we see them
doing.

So much by way of preface. Now let us see
what our author has to say

To begin with, he, like many another before
and since, found it disagreeable to be wakened in
the morning. When a person is enjoying sweet
repose in his bed, to be suddenly awakened by the
rude, blustering voice of a vociferous ostler was
distinctly annoying. More annoying still, however,
to lose the coach ; and so there was no help for
it, provided the stage was to be caught. The
morning was very fine when the passengers,
thus untimely roused, entered the coach Nature
smiled around them, who only yawned in her face
in return. Pity, thought our author, that they were

not to ride on horseback · they could then enjoy
the pleasures of the morning, snuff the perfumes
of the fields, hear the music of the grove and
the concert of the wood.

These reflections were cut short by the crossing
of the Tyne by ferry. The bridge had fallen on
November 17th, 1771, and the temporary ferry
established from the Swirl, Sandgate, to the
south shore was the source of much inconvenience
and delay. The coach was put across on a raft
or barge, but in directing operations to that end,
the ferryman was not to be hurried. One had to
wait the pleasure of that arbitrary little Bashaw,
who would not move beyond the rule of his
own authority, or mitigate the sentence of those
who were condemned to travel in a stage-coach
within a ferry-boat.

Our author, as he hated every idea of slavery
and oppression, was not a little offended at the
expressions of authority used on this occasion by
the august legislator of the ferry. The passengers
were now in the barge, and obliged to sit quiet
until this tyrant gave orders for departure. The
vehicle for carrying coach and passengers across
the river was the most tiresome and heavy that
ever was invented. Four rowers in a small boat
dragged the ponderous ferry across the river,
very slowly and with great exertions, and almost an
hour was consumed in thus breasting the yellow
current of the broad and swiftly-running Tyne.
Meanwhile, there was plenty of time to reflect on
what might happen on the passage, and abundant

opportunity for putting up a few ejaculations to Heaven to preserve them all from the dangers of ferryboats and tyrants

But the voyage at last came to an end. So soon as they were landed on the south side of the river Tyne, they were saluted by a blackbird, who welcomed them to the county of Durham. It seemed to take pleasure in seeing them fairly out of the domains of Charon, and whistled cheerfully on their arrival. "Nature," said Mr. James Murray to himself, "is the mistress of real pleasure. this same blackbird cannot suffer us to pass by without contributing to our happiness. Liberty (he continued) seems to be the first principle of music. Slaves can never sing from the heart "

No: they sing, like everyone else, from the throat.

But these observations carried them beyond Gateshead and to the ascent of the Fell, along whose steep sides the pleasures of the morning increased upon them. The whins and briars sent forth a fragrance exceedingly delightful, and on every side of the coach peerless drops of dew hung dangling upon the blossoms of the thorns, adding to the perfume Aurora now began to streak the western sky—something wrong with the solar system that morning, for the sun commonly rises in the east—and the spangled heavens announced the advent of the King of Day. Sol at last appeared, and spread his healthful beams over the hills and valleys, and the wild beasts now

retired to their dens, and those timorous animals
that go abroad in the night to seek their food were
also withdrawn to the thickets. The hares, as
an exception—and yet this was not the lunatic
month of March—were skipping across the lawns,
tasting the dewy glade for their morning's repast.
The skylark was skylarking—or, rather, was
already mounted on high, serenading his dame
with mirthful glee and pleasure (Here follow
two pages of moral reflections on skylarks and
fashionable debauchees, with conclusions in favour
of the larks, and severe condemnation of "libidinous
children of licentiousness," who are bidden " go
to the lark, ye slaves of pollution, and be wise.
He does not stroll through the grove or thicket to
search for some new amour, but keeps strictly to
the ties of conjugal affection, and cherishes the
partner of his natural concerns.")

In the midst of these idyllic contemplations,
a grave and solemn scene opened to the view.
Hazlett, who had robbed the mail in 1770, hung
on a gibbet at the left hand. "Unfortunate and
infatuated Hazlett! Hadst thou robbed the nation
of millions, instead of robbing the mail and
pilfering a few shillings from a testy old maid,
thou hadst not been hanging, a spectacle to
passers-by and a prey to crows Thy case was
pitiable—but there was no mercy: thou wast
poor, and thy sin unpardonable. Hadst thou
robbed to support the Crown, and murdered for
the Monarchy, thou might'st have been yet alive."

The place where Hazlett hung, the writer

considered to be the finest place in the world for a ghost-walk. "At the foot of a wild romantic mountain, near the side of a small lake, are his remains; his shadow appears in the water and suggests the idea of two malefactors. The imagination may easily conjure up his ghost. Many spirits have been seen in wilds not so fit for the purpose. This robber is perhaps the genius of the Fell, and walks in the gloomy shades of night by the side of this little lake. This (he adds—it must have been a truly comforting thought to the other passengers) is all supposition." The dreary place was one well calculated for raising gloomy ideas, tending to craze the imagination

After this, it was a relief to reach Durham, a very picturesquely situated city with a grand cathedral and bishop's palace. The pleasant banks on the west side of the river Wear were adorned with stately trees, mingled with shrubs of various kinds, which brought to one's mind the romantic ideas of ancient story, when swains and nymphs sang their loves amongst trees by the side of some enchanted river. The abbey and the castle called to mind those enchanted places where knights-errant were confined for many years, until delivered by some friend who knew how to dissolve the chains and charm the necromancy.

Durham, he thought, would be a very fine place, were it not for the swarms of clergy in it, who devoured every extensive living without being of any real service to the public. The common people in Durham were very ignorant and great profaners of the Sabbath Day, and, indeed, over almost the whole of England the greatest ignorance and vice were under the noses of the bishops. He would not pretend to give a reason for this, but the fact was apparent.

Durham was a very healthful place—the soil dry, the air wholesome; but the Cathedral dignitaries performed worship rather as a grievous task than as a matter of choice, a thing not infrequently to be observed in our own days. The woman who showed the shrine of St. Cuthbert did not understand Mr. Murray when he referred to the Resurrection, a fact that gave him a good opportunity to enlarge upon the practically heathenish state of Durham's ecclesiastical surroundings.

All this sightseeing, and these reflections and observations at Durham (and a good many more from which the reader shall be spared) were rendered possible by a lengthy halt made by the coach in that city. Thus there was ample time for seeing the cathedral—"very noble and delightful to the eyes of those who had a taste for antiquity or Gothic magnificence," he says.

After they were wearied with sauntering in this old Gothic abbey, they went down to the river side. There the person who was fond of rural pleasures might riot at large. Comparisons drawn on the spot between the choristers of the grove, who sang from the heart, and the minor canons and prebendaries of the cathedral, who wearily performed their duties for a living, were, naturally, greatly to the disadvantage of the dignified clergy.

Strolling through the suburb of Old Elvet, the company at last returned to the inn—the " New Inn " it was called. The landlord of this hostelry was a jolly, honest man ; his house spacious, and fit even to serve the Bishop All things were cheap, good, and clean at this inn. If a person came in well pleased, he would find nothing to offend him, provided he did not create some offence to himself—which sounds just a little confused.

While our itinerant chronicler was noting down all these things, orders were given for departure, and so he had hurriedly to conclude.

And now, turning from wayside reflections, we get a description of the passengers. The coach, when it left Newcastle, was full. Four ladies, a gentleman of the sword and our humble servant made up its principal contents. They sat in silence for some time, until they were jolted into good humour by the motion of the vehicle, which opened their several social faculties. One of their female companions, who was a North Briton, a jolly, middle-aged matron with abundance of good sense and humour, entertained the company for a quarter of an hour with the history of her travels. She had made the tour of Europe, and had visited the most remarkable places in Christendom, in the quality of a dutiful wife, attending her valetudinary husband, travelling for the recovery of his health. Her easy, unaffected manner in telling a story made her exceedingly good company, and none had the least inclination to interrupt her until she was pleased to cease. She knew how to time her discourse, and never, like the generality of her sex, degenerated into tediousness and insipidity.

At every stage she was a conformist to all the measures of the company, and went into every social proposal that was made.

Another companion was a widow lady of Newcastle, quite as agreeable as the former. She understood how to make them laugh. Unfortunately, she only went one stage, and they then lost the pleasure of her company.

The third passenger was a Newcastle lady, well known in the literary world for her useful performances for the benefit of youth. This female triumvirate would have been much upon a par had they all been travellers, for their gifts of conversation were much alike; but the lady who had taken the tour of Europe possessed in that the advantage of circumstances.

The fourth lady was the Scottish lady's servant. As she said nothing the whole way (remarks Mr. Murray), I shall say nothing of her.

The fifth person was an officer in the army, who appeared very drowsy in the morning, and came forth of his chamber with every appearance of reluctance. His hair was dishevelled and quite out of queue, and he seemed to be as ready for a sleep as if he had not been to bed. He was, for a time, as dumb as a Quaker when not moved by the spirit, and by continuing in silence, at last fell asleep until they had completed nearly half the first stage. During this time, Mr. Murray sarcastically observes, he said no ill

They finished their first stage without exchanging many words with this son of Mars, except some of those flimsy compliments gentlemen of the sword pay frequently to the ladies. After a dish of warm tea the tissues of his tongue were loosed, and he began to let his companions know that he was an officer in the army, and a man of some consequence. He seemed to be very fond of war, and spoke in high terms upon the usefulness of a standing army. When he had exhausted his whole fund of military arguments in favour of slavery and oppression, Mr Murray observed to him that a standing army had a bad appearance in a free country, and put it in the power of the Crown to enslave the nation—with the like arguments, continued for an unconscionable space.

It is not at all surprising that the soldier resented this. The spirit of Mars began to work within him, and he threatened that if he were near a Justice of the Peace he would have this argumentative person fined for hindering him from getting recruits, adding that he once had a man fined for persuading others not to enlist in his Majesty's service.

To this Mr. Murray rejoined that the officer

certainly had a right to say all the fine things
he could to recommend the service of his master,
but, having done that, he had no more to do;
and that any man had also a right to tell his
friends, whom he saw ready to be seduced into
bondage, that they were born free, and ought to
take care how they gave up their liberty—
together with remarks derogatory of the justice
of courts martial.

Our author did not, however, find this military
hero a bloodthirsty man, for, by his own confes-
sion, he and a brother officer had a few months
before surrendered their purses to a highwayman
between London and Highgate for fear of blood-
shed. This showed that some officers were abundantly peaceable in time of danger, and discovered
no inclination for taking people's lives. This
gentleman of sword and pistol, in particular, had
a great many solid reasons why men should not
adventure their lives for a little money. He said
there was no courage in fighting a highwayman,
and no honour to be had in the victory over one;
that soldiers should preserve their lives for the
service of the country in case of war, and not run
the risk of losing them by foolish adventure.

These reasons did not altogether satisfy the
ladies, for one of them observed that robbers were
at war alike with laws and governments, and that
the King's servants were hired to keep the peace
and to defend the King's subjects from violence;
that officers in the army were as much obliged by
their office and character to fight robbers as they

were bound to fight the French, or any other
enemy, and that footpads were invaders of the
people's rights and properties, and ought to be
resisted by men whose profession it was to fight,
and who were well paid for so doing. It was for
money all the officers in the army served the
King and fought his battles, and why should they
not as well fight for money in a stage-coach as
in a castle or a field? She insisted that only one
of them could have been killed by the highwayman, or perhaps but wounded, and there were
several chances that he might have missed them
both. But, supposing the worst—that one had
been shot—it was only the chance of war, and
the other might have secured the robber, which
would have been of more service to the country
than the life of the officer. In short, she observed,
it had the appearance more of cowardice than
disregard for money, for two officers to surrender
their purses to a single highwayman, who had
nothing but one pistol.

The lady's reflections were severely felt by
the young swordsman, and produced a solemn
silence in the coach for a quarter of an hour,
during which time some fell asleep, and so continued until coming to the next inn, where the
horses were changed There two or three glasses
of port restored the officer's courage, and he
determined, in case of an attack, to defend every
one from the assaults of all highwaymen what-
soever. To show the courage that sometimes
animated him, he told the story of how he had

dealt with a starving mob in Dumfries. The
hungry people of that town, not disposed to perish
while food was abundant, and corn held by the
farmers and corn-factors for higher prices, assembled to protest against such methods; and the
magistrates, who thought the people had a right
to starve, sent for the military to oblige them to
famish discreetly or else be shot Our hero had
command of the party, where, according to his
own testimony, he performed wonders. The poor
people were shot like woodcocks, and those who
could get away with safety were glad to return
home to wrestle with hunger until Heaven should
think fit to provide for them.

The officer was very liberal in abusing those
whom he called "the mob," and said they were
ignorant, obstinate and wicked, and added that he
thought it no crime to destroy hundreds of them.

The lady who had already given him a lecture
then began to put him in mind of the footpad
whom he and his brother officer had suffered to
escape with their purses, and asked him how he
would quell a number of highwaymen. Taken off
his guard at the mention of footpads, he stared
out of the window with a sort of wildness, as if
one had been at the coach door.

Nothing was seen worthy of note until the
coach came to Grantham, which place they
reached about seven in the evening. The first
things, remarks Mr Murray—with all the
air of a profound and interesting discovery—
that travellers saw in approaching large towns

were, generally speaking, the church steeples. Ordinarily higher than the rest of the buildings, they were—remarkable to relate—on that account the more conspicuous. The steeple of Grantham was pretty high, and saluted one's eyes at a good distance before the town was approached. It seemed to be of the pyramidical kind.

Grantham was a pleasant place, although the houses were indifferently built. On reaching it, they wandered through the town before returning to the inn for supper, when the captain took care to say some civil things to the landlady's sister, who was a very handsome young woman. It was, however, easy to perceive that she was acquainted with these civilities, and could distinguish between truth and falsehood. She made the captain keep his distance in such a manner as put an entire end to his compliments. The fineness of her person and the beauty of her complexion were joined with a modest severity that protected her from the rudeness and insults which gentlemen think themselves entitled to use towards a chamber-maid, the character she acted in.

After supper was done, the coach-party were informed that some of Mr Garrick's servants were that night to exhibit in an old thatched house in a corner of the town They had come abroad into the country during the summer vacation, to see if they could find anything to keep their grinders going until the opening of Drury Lane Theatre. They were that night to play the *West Indian* and the *Jubilee*.

The whole of the passengers went to see the performance. The actors played their parts very indifferently, but, after all, not so badly but that one could, with some trouble, manage to perceive as much meaning in their actions as to be able to distinguish between an honest man and a rogue. Our ingenious and imaginative Mr Murray thought it must be dangerous for an actor to play the rogue often, for fear of his performance becoming something more than an imitation. But after all, he says, with the fine intolerant scorn of the old-time dissenting minister, the generality of players had little morality to lose.

It was a very poor theatre—being, indeed, not a theatre at all, and little better than a barn The audience, however, was good, and well dressed, and the ladies handsome. The performance was over by eleven o'clock, and the company dismissed. Mr. Murray concludes his account of the evening's entertainment by very sourly observing that their time and that of the rest of the audience might have been better employed than in seeing a few stupid rogues endeavouring to imitate what some of them really were.

The coach left Grantham at two o'clock the next morning; much too early, considering the short rest the night's gaiety had left them. But there was no choice—they were under authority, and had to obey. That person would be a fool who, having paid £3 8s. 6d. for a seat in a stage-

coach from Newcastle to London, should consent
to lose it by not rising betimes. The worst of
it was, that here one had to take care of one's
self, because no one would wait upon him or
return him his money. Observe the passengers,
therefore, all, in the coach by 2 a.m. The company being seated, the driver went off as fast
as if he would have driven them to Stamford in
the twinkling of an eye. So early was the hour
that we are not surprised to be told that the author
fell asleep by the time they were clear of the
town, and doubtless the others did the same.
It may be remarked here that a very excellent
proof of this being a fictitious journey is found
in there being no mention of the passengers being
turned out of the coach to walk up the steep
Spitalgate Hill—a thing always necessary at
that period of coaching history

The remainder of this not-inaptly named
Travels of the Imagination is made up chiefly
of a long disquisition upon sleep—itself highly
soporific—which only gives place to remarks
upon the journey when the coach arrives on
Highgate Hill. Coming over that eminence,
they had a peep at London.

"It must be a wonderful holy place," he
suggests to the other passengers, "there are so
many church steeples to be seen."

The others, who must have known better, said
nothing.

"Are we there?" he asked when they had
reached Islington.

"No, not there yet."

"Is it a large place: four times as large, for instance, as Newcastle?"

"Ten times as large"

"Where are the town walls?"

"There are no walls."

At last they reached Holborn, and the end of the journey, where the company dispersed and our chronicler went to bed.

CHAPTER III

DOWN THE ROAD IN DAYS OF YORE

II—From London to Newcastle-on-Tyne in 1830

WE also will make a tour down the road. It shall not be, in the strictly accurate sense of the word, a "journey," for we shall travel continuously by night as well as day—a thing quite unknown when that word was first brought into use, and unknown to coaching until about 1780, when coaches first began to go both day and night, instead of inning at sundown at some convenient hostelry on the road.

It matters little what road we take, but as Mr. Murray came to town from Newcastle, we may as well pay a return visit along that same highway—the Great North Road. He does not explain how he came through Highgate, but for our part, the first sixty miles or so go along the Old North Road, and we do not touch Highgate at all.

Now, since we are setting out merely for the purpose of seeing something of what life is like on a great highway, there is no need to mortify the flesh by arising early in the blushing hours of dawn, to the tune of the watchman's cry of

"five o'clock and a fine morning!" and so we
will e'en, like Christians and Britons able to
call their souls their own, go by the afternoon
coach Let the "Lord Nelson" in this year
1830 go if it will from the "Saracen's Head,"
Snow Hill, at half-past six in the morning. For
ourselves, we will wait until a quarter to three
in the afternoon, and take the "Lord Wellington"
from the "Bull and Mouth." We can do no
better, for the "Lord Wellington" goes the 274
miles in 30 hours, which a simple calculation
resolves into 9 miles an hour, including stops.
The fare to Newcastle is £5 15s. inside, or about
5d. a mile. Outside, it is £3 10s., or a fraction
over 3d. a mile. As our trip is taken in summer-
time, we will go outside; and so, although a good
deal of the journey will have to be through the
night, we, at least, shall not have the disadvantage of being stewed during the daytime in
the intolerable atmosphere of the inside of a
stage-coach on a July day. Why, indeed, coach-
proprietors do not themselves observe that in
summer-time the outside is the most desirable
place, and charge accordingly, is not easily understood; nor, indeed, to be understood at all. That
clever fellow De Quincey notices this, and points
out that, although the roof is generally regarded
by passengers and everyone else connected
with coaching as the attic, and the inside
as the drawing-room, only to be tenanted by
gentlefolk, the inside is really the coal-cellar in
disguise.

We recollect, being old travellers, that the fares to Newcastle used to be much cheaper Time was when they were only four guineas inside and £2 10s. outside, but prices went up during the late wars with France, and they have stayed up ever since. The travelling, however, is better by some five hours than it was fifteen years ago.

Here we are at the " Bull and Mouth,' in St. Martin's-le-Grand, now newly rebuilt by Sherman, and named the " Queen's." It is a handsome building of red brick, with Portland stone dressings, but the old stables are still to be seen at the side, in Bull-and-Mouth Street. A strong and penetrating aroma of horses and straw pervades the neighbourhood.

Wonderful building, the new General Post Office, opened last year, nearly opposite. They say the Government has got something very like a white elephant in that vast pile. A great deal too big for present needs, or, indeed, for any possible extension of Post Office business. Here's the " Lord Wellington." What's that the yard-porter says?—He says " they don't call it nothin' but the 'Vellington' now."—Smart turn-out, is it not, with its yellow wheels and body to match? You can tell Sherman's coaches any-where by that colouring What a d——d nuisance those boys are, pestering one to buy things one doesn't want! No, be off with you, we don't want any braces or pocket looking-glasses, nor the " Life and Portrait of His Late

THE "QUEEN'S HOTEL" AND GENERAL POST OFFICE.

After T. Allom.

Majesty," nor any "Sure Cure for Fleas"—use it on yourselves, you dirty-looking devils!

Thank goodness! we're off, and the sooner we're out of this traffic and off the stones at Kingsland Turnpike the better These paved streets are so noisy, one can scarcely hear oneself talk, and the rattling sends a jar up one's spine. How London grows! we shall soon see the houses stretching past Kingsland and swallowing up the country lanes of Dalston and Stoke Newington.

Hal-lo! That was a near shave. Confound those brewers' drays; Shoreditch is always full of 'em; might have sent us slap over Why don't you keep your eyes open, fool?

The drayman offers to fight us all, one after the other, with one hand tied behind his back, for sixpence a head, money down; but though we have some of "the Fancy" aboard, the "Wellington" can't stop for a mill in the middle of Shoreditch High Street.

Now at last we're fairly in the country. If you look back you'll be able to see St. Paul's. This is Stamford Hill, where the rich City indigo and East and West India merchants live. Warm men, all of them. There, ahead of us, on the right, goes the river Lea: as pretty fishing there as you'd find even in the famous trout streams of Hampshire. What a quaint, quiet rural place this is at Tottenham! And Edmonton, with its tea-gardens; why, London might be fifty miles away!

Here we are, already at Waltham Cross, and

at our first change. This is something like travel-
ing! We change horses at the "Falcon" in little
more than two minutes, and so are off again,
on the ten-mile stage to Ware, through the long
narrow street of Cheshunt, past the New River
at Broxbourne, and along the broad thoroughfare
of Hoddesdon. At Ware we change teams at the
"Saracen's Head," and four fine strong-limbed
chestnuts are put in, to take us the rather hilly
stage on to Buntingford. At this sleepy little
town they take care to give us as strong a team as
you will find in any coach on any road, for the
way rises steadily for some miles over Royston
Downs. A good thing for the horses that the
stage on to Royston town is not more than seven
miles "I believe you, sir," says the coachman;
"vy, I've heerd my father say, vot driv' over this
'ere road thirty year ago, that he vore out many
a good 'orse on this stage; an' 'e vere a careful
man too, as you might say, and turned out every
blessed one, *h*inside or *h*out, to valk up-hill for
two mile, wet or fine; strike me blue if he didn't."

"They talk of lowering the road through the
top of Reed Hill, don't they, coachman?"

"Oh! yes; they torks, and that's about all
they does do. Lot o' good torking does my 'orses.
Vot *I* vants to know is, v'y does we pay the
turnpikes?"

We change at the "Red Lion," Royston, and
then come on to the galloping ground that brings
us smartly, along a level road, to Arrington Bridge,
the spelling of whose name is a matter of divergent

opinions among the compilers of road-books. But whether called Arrington or Harrington, it is a pretty, retired spot, with a handsome inn and an equally handsome row of houses opposite.

"Will you please to alight?" asks the stately landlady of the "Hardwicke Arms" inn and posting-house, with perhaps a little too much air of condescension towards us, as coach-passengers. We of the stage-coaches—nay, even those of the mails—occupy only a second place in the consideration of mine host and hostess of this, one of the finest inns on the road. Their posting business brings them some very free-handed customers, and their position, hard by my lord of Hardwicke's grand seat of Wimpole, spoils them for mere ordinary everyday folks.

However, it is now more than half-past seven o'clock, and we have had no bite nor sup since two. Therefore we alight at the landlady's bidding and hasten into the inn, to make as good a supper as possible in the twenty minutes allowed.

Half a crown each, in all conscience, for two cups of tea, and some bread and butter, cold ham and eggs! We climb up to our places, dissatisfied. Soon the quiet spot falls away behind, as our horses get into their stride; and as we leave, so does a yellow po'shay dash up, and convert the apparently sleepy knot of smocked post-boys and shirt-sleeved ostlers, who have been lounging about the stable entrance, into a very alert and wide-awake throng.

Caxton, a busy thoroughfare village, where the

great " George " inn does a very large business, is passed, and soon, along this flattest of flat roads, that grim relic, Caxton Gibbet, rises dark and forbidding against the translucent evening sky. Does the troubled ghost of young Gatward, gibbeted here eighty years ago for robbing the mail on this lonely spot, ever revisit the scene, we wonder?

The wise, inscrutable stars hang trembling in the sky, and the sickle moon is shining softly, as, having passed Papworth St Everard, we drop gently down through Godmanchester and draw up in front of the "George" at Huntingdon, 58½ miles from London, at ten o'clock.

We take the opportunity afforded by the change of filling our pocket-flasks with some rich brown brandy of the right sort, and invest in some of those very special veal-and-ham sandwiches for which good Mrs Ekin has been famous these years past. Our coachman "leaves us here," and we tip him eighteenpence apiece when he comes round to inform us of the fact.

The new coachman, after some little conversation with the outgoing incumbent of the bench, in which we catch the remark made to the newcomer that some articles or some persons are "a pretty fair lot, taking 'em all round"—a criticism that evidently sizes us up for the benefit of his *confrère*—climbs into his seat, and giving us all a comprehensive and impartial glance, settles himself down comfortably. "All right, Tom?" he asks the guard over his shoulder,

" Yes," answers that functionary. " Then give
'em their heads, Bill," he says to the ostler ; and
away we go into the moonlit night at a steady
pace.

The box-seat passenger, who very successfully
kept the original coachman in conversation nearly
all the way from London to Huntingdon, does not
seem to quite hit it off with our new whip, who
is inclined to be huffish, or, at the least of it, given
to silence and keeping his own counsel. " Have
a weed, coachman ? " he asks, after some in-
effectual attempts to get more than a grunt out
of him. " Don't mind if I do," is the ungracious
reply, and he takes the proffered cigar and—puts
it into some pocket somewhere beneath the
voluminous capes of his greatcoat. After this,
silence reigns supreme For ourselves, we have
chatted throughout the day, and now begin to feel
—not sleepy, but meditative.

The moon now rides in unsullied glory through
the azure sky. We top Alconbury Hill at a few
minutes to twelve, and come to the junction of
the Old North and the Great North Roads.
Everything stands out as clearly as if it were
daylight, but with a certain ghost-like and un-
canny effect. " The obelisk," as the coachmen
have learned to call the great milestone at the
junction of the roads (it is really a square
pedestal) looks particularly spectral, but is not
the airy nothing it seems—as the coachman on
the Edinburgh Mail discovered, a little while
ago. The guard tells us all about it. The usual

thing. Too much to drink at the hospitable bar of the "George," at Huntingdon, and a doubt as to which of the two milestones he saw, on coming up the road, was the real one. The guard and all the outsides were in similar case—it was Christmas, and men made merry—and so there was nothing for it but to try their quality. Unfortunately, he drove into the real stone, and not its spectral duplicate, conjured up by the effects of strong liquors. We see the broken railings and the dismounted stone ball that once capped the thing as we pass. The local surgeon mended the resultant broken limbs at the "Wheatsheaf," whose lighted windows fall into our wake as we commence the descent of Stonegate Hill.

Stilton. By this time we are too drowsy to note whether we changed at the "Bell" or at its rival, directly opposite, the "Angel." At any rate, nobody asks us if we would not like a nice real Stilton cheese to take with us, as they usually do: it is midnight.

We now pass Norman Cross, and come in another eight miles to Wansford turnpike, where the gate is closed and the pikeman gone to bed. "Blow up for the gate," said the coachman, when we were drawing near, to the guard, who blew his horn accordingly; but it does not seem to have disturbed the dreams of the janitor. "Gate, gate!" cry the guard and coachman in stentorian chorus. The guard himself descends, and blows a furious series of blasts in the doorway, while the coachman lashes the casement windows.

THE TURNPIKE GATE.

From a contemporary lithograph.

At last a shuffling and fumbling are heard within, and the door is opened. The pikeman has not been to bed after all; he was, and is, only drunk, and had fallen into a sottish sleep. He now opens the gate, in the midst of much disinterested advice from both our officials—the guard advising him to stick to Old Tom and leave brandy alone, and the coachman pointing out that the Mail will be down presently and that he had better leave the gate open if he does not wish to present the Postmaster-General with forty shillings, that being the penalty to which a pike-keeper is liable who does not leave a clear passage for His Majesty's Mails.

We now cross Wansford Bridge, a very long and narrow stone structure over the river Nene. Having done so, slowly and with caution, we know no more: sleep descends insensibly upon us.

. . . Immeasurable æons of time pass by. We are floating with rhythmic wings in the pure ether of some unterrestrial paradise. Our gross earthly integument (twelve stone and a few extra pounds avoirdupois of flesh and blood and bone) has fallen away. We want nothing to eat, for ever and ever, and have left everything gross and unspiritual far, far below us, and . . . . a fearful crash! Convulsively, instinctively, our arms are thrown out, and we awake, tenaciously grasping one another. What is this that has brought us down to earth again and made us unwillingly assume once more that corporeal

hundredweight, or thereabouts, we had left so
gladly behind? Are we overturned?

No; it was nothing: nothing, that is to say,
but the hunchbacked bridge over the river
Welland, that leads from Stamford Baron into
Stamford Town. It is only the customary bump
and lurch, the guard informs us. May all archi-
tects of hunchback bridges be converted from
straight-backed human beings into bowed and
crooked likenesses of their own abominable
creations! We will keep awake, lest another
such rude awaking await us.

With this intent we gaze, wide-eyed, upon
Stamford Town, its noble buildings wrapped
round in midnight quiet, the moon shining
here full upon the mullioned stone windows of
some ancient mansion, there casting impenetrable
black shadows, making dark mysteries of grand
architectual doorways decorated with curious
scutcheons and overhung with heavy pediments,
like beetling eyebrows Grand churches whose
spires soar away, away far into the sky, astonish
our newly-awakened vision as the coachman carefully guides the coach through the narrow and
crooked streets, in which the shadows from
cornices and roof-tops lie so black and sharp
that none but he who has driven here before could
surely bring this coach safely through. Once or
twice we have quailed as he has driven straight
at some solid wall, and have breathed again
when it has proved to be only some oblique
monstrous silhouetted image cast athwart the

way. Fear only leaves us when we are clear
of the town and once more on the unobstructed
road ; then only is there leisure for the mind to
dwell upon the beauties of that glorious old
stone-built town. We are thus ruminating when,
between Great Casterton and Stretton, where
we enter Rutlandshire, the glaring lamps of a
swiftly approaching coach lurch forward out of
the long perspective of road, and, with a clatter
of harness and a sharp crunching of wheels, fall
away, as in a vision. The guard, answering some
one's question, says it is the Leeds " Rocking-
ham," due in London at something after ten in
the morning.

The determination to keep awake was heroic,
but without avail. Even the screaming and
grumbling of the skid and the straining of the
wheels down Spitalgate Hill into Grantham did
not suffice to quite waken us. But what that
noise and the jarring of the wheels failed to do, the
stoppage at the " George " at Grantham and the
sudden quiet *do* succeed in. Our friend the moon
has by now sunk to rest, and a pallid dawn has
come, someone remarks that it is past three
o'clock in the morning, and someone else is
wakened and hauled forth from amid the snoring
insides, whose snores become gasps and gulps, and
then resolve themselves into the yawns and peevish
exclamations of tired men. The person thus
awakened proves to be a passenger who had booked
to Colsterworth, which is a little village we have
now left eight miles behind us. He had been

asleep), and as Colsterworth is not one of our stopping or changing places, the guard forgot all about him until the change at Grantham. The passenger and the guard are now waging a furious war of words on the resounding pavements of the sleeping town. It seems that the unfortunate inside, besides being himself carried so far beyond his destination, has a heavy portmanteau in the like predicament. If he had been a little bigger and the guard a little smaller, his fury would perhaps make him fall upon that official and personally chastise him. As it is, he resorts to abuse. Windows of surrounding houses now begin to be thrown up, and nightcapped heads to inquire " what the d——l 's the matter, and if it can't be settled somewhere else or at some more convenient season ? " The guard says " This 'ere gent wot's abusing of me like a blooming pickpocket goes to sleep and gets kerried past where he wants to get out, and when I pulls him out, 'stead of taking 'im him on to Newark or York, 'e ——" " Shut up," exclaims a fierce voice from above· " can't a man get a wink of sleep for you fellows ? "

So, the change being put to, the altercation is concluded in undertones, and we roll off ; the irate passenger to bed at the " George," vowing he will get a legal remedy against the proprietors of the " Wellington " for the unheard-of outrage.

At Newark, a hundred and twenty-five miles of our journey performed, it is broad daylight as the coach rolls, making the echoes resound, into the great market-square. Clock-faces—a little

blanched and debauched-looking to our fancy—
proclaim the hour to be 5.30 a.m. The change
is waiting for us in front of the "Saracen's Head,"
and so is our new coachman. The old one leaves
us, but before doing so "kicks us"—as the ex-
pressive phraseology of the road has it—for the
usual fees. He has been, so far as we remember
him, a dour, silent, unsociable man, but we think
that, perhaps, as we have been asleep during
the best part of his reign on the box-seat, any
qualities he may possess have not had their due
opportunity, and so he gets two shillings from
ourselves. A passenger behind us gives him a
shilling, which he promptly spits on and turns,
" for luck " as he says, and " in 'opes it'll grow."
The passenger who gave it him says, thereupon—
in a broad Scots accent—that he is "an impudent
fellow, and desairves to get nothing at all ;" to
which the jarvey rejoins that he has in his time
brought many a Scotchman from Scotland, but,
" this is the fust time, blow me, that *hever* I see
one agoin' back !"—which is a very dark and
mysterious saying. What did he mean ?

Our new coachman is a complete change from
our late Jehu. He is a spruce, cheerful fellow,
neat and well brushed, youthful and prepossessing.
" Good morning, gentlemen," he says cheerily.
"another fine day." We had not noticed it. All
we had observed was of each other, and that as
every other looked pale, wearied and heavy-eyed,
so we rightly judged must be our own condition.

" Chk ! " says our youthful charioteer to his

horses, and away they bound. Newark market-square glides by, and we are crossing the Trent, over a long bridge. "Newark Castle, gentlemen," says our coachman, jerking his whip to the left hand; and there we see, rising from the banks of the broad river, the crumbling, time-stained towers of a ruined mediæval fortress. Much he has to say of it, for he is intelligent beyond the ordinary run. A good and graceful whip, too—one of the new school . much persuasion and little punishment for the horses, who certainly seem to put forward their best paces at his merest suggestion. It is a good, flat, and fairly straight road, this ten-mile stage to Scarthing Moor. We cross the Trent again, then a low-lying tract of water-meadows, where the night mists still cling in ghost-like wisps to the grass, and then several small villages "This"—says our coachman, pointing to a church beside the road, and down the street of one of these little villages—"this is where Oliver Cromwell came from."

"What is the name of it?" we ask, knowing that, whatever its name, the Protector came from quite a different place.

"Cromwell," he says

So this was probably the original seat of that family many centuries before Oliver came into the world, which has since then been so greatly exercised about him.

"Blow up for the change," says the coachman to the guard, as, having passed through Carlton-on-Trent, Sutton-on-Trent, and round the

awkward bend of the road at Weston, we approach
Scarthing Moor and the " Black Bull." " They're
a sleepy lot at the ' Bull,' " he says, in explanation.
The guard produces the " yard of tin " from the
horn-basket, and sounds a melodious tantara. quite
unnecessarily, after all, it seems, for, quite a
distance off, the ostler, dressed after his kind
in trousers and shirt only, with braces dangling
about him, is seen standing in the road, with
the change ready and waiting.

"Got up before you found yourself, this
morning ? " asks the coachman.

The ostler says he don't take no sauce from
no boys what ain't been breeched above a twelve-
month

"All right, Sam," replies the coachman; " your
'art's all right, if you *have* got a 'ed full of
wool. Shouldn't wonder if you don't make up for
this mistake of yourn by sleepin' it out for a
month of Sundays after this If so be you do,
jest hang the keys of the stable outside, and when
we come down agen, Jim and me 'll put 'em in
ourselves, won't we, Jim ? "

Jim says they will, and will petition Guv'ment
to pension him off, and retire him to the " R'yal
'Orsepital for Towheads."

Evidently some ancient feud between the
ostler and the coach is in progress, and still far
from being settled. The ostler sulkily watches
us out of sight, as we make our next stage to
Retford. The clocks in the market-place of that
busy little town mark half-past seven, and the

"White Hart," where we drop a passenger for
the Gainsborough coach and another for Chester-
field, and take up another for York, is a busy
scene Appetising aromas of early breakfasts being
prepared put a keener edge upon our already
sharpened appetites, and we all devoutly wish we
were at Doncaster, where *our* breakfast awaits the
coming of the coach Across Barnby Moor, past
the great "Bell" inn, we take our way, and come
to one more change, at the "Crown," Bawtry;
then hie away for Doncaster, which we reach,
past Rossington Bridge and the famous St Leger
course, at half-past nine o'clock.

"Twenty minutes for breakfast, gentlemen,"
announces the coachman as we pull up in front of
the "New Angel" inn; while the guard, who has
come with us all the way from London, now
announces that he goes no farther. We give him
half a crown, and hasten, as well as stiffened limbs
allow, down the ladder placed for us outsides to
alight by, to the breakfast-room.

We catch a glimpse of ourselves in a mirror
as we enter. Heavens! is it possible an all-night
journey can make so great a difference in a man's
personal appearance? While here is a lady who
has been an inside passenger all the way from
town, and yet looks as fresh and blooming as
though she had but just dressed for a walk How
do they manage it, those delicate creatures?

Our friend, who says he is starving, refuses
to discuss this question. He remarks, with eye
wildly roving o'er the well-laden table-cloth,

that something to eat and drink is more to the point. We cannot gainsay the contention, and do not attempt it, but sink into a chair

"Coffee, sir; tea, sir; 'ot roll; 'am and heggs Yorkshire brawn, tongue," suggests the waiter, swiftly.

We select something and fall-to. After all, it is worth while to take a long coach journey, even if it be only for the appetite it gives one. Here we are, all of us, eating and drinking as though we had taken no meals for a week past. Yes, another cup of coffee, please, and I'll thank you to pass the——

"Time's up, gents; coach just agoin' to start!"

"Oh! here, I say, you know. We've only just sat down"

"Ain't got more'n 'nother couple o' minutes," says the new guard, and so, appetite not fully satisfied, we all troop out and resume our places.

Our coach goes the hilly route, by Ferrybridge and Tadcaster, to York. We change on the short stage out of Doncaster, at Robin Hood's Well, where the rival inns, the "New" and the "Robin Hood," occupy opposite sides of the road; and again at Ferrybridge, at the "Swan," where our smart coachman resigns his seat to an enormously fat man, weighing nearly, if not quite, twenty stone. He is so unwieldy that quite a number of the "Swan" postboys gather round him, and by dint of much sustained effort, do at last succeed in pulling and pushing him into his place, resembling in so doing the Liliputians manipulating Gulliver; the coachman himself, breathing like a grampus, encouraging them by calling out, " That's it, lads, another heave like t'last does it All together again, and I'll mak' it a gallon ! "

Across the river Aire to Brotherton, and thence through Sherburn to Tadcaster, where, having changed at the " White Horse," we come along a level stage into York; the new guard, who rejoices in the possession of a key-bugle and a good ear for music, signalising our entrance by playing, in excellent style, " The Days when we went Gipsying, a Long Time Ago."

The coach dines at York. The " Black Swan," to which we come, is a house historic in the annals of coaching, for it was from its door that the original York and London stage set forth ; but it is a very plain and heavy building. Half an hour is allowed for dining, and, unlike the majority of houses down the road, the table-cloth and the knives and forks and glasses are *not* the only things in readiness.

" What have you got, waiter ? "

" Hot roast beef, sir, just coming in ; very prime."

" Haven't you any cold chicken for a lady here ? "

" Yessir; cold chicken on the table, sir ; in front of you, sir."

" You call *that* chicken, waiter ! why, it's only a skeleton. Take it away and give it to the dog in the yard."

"Very sorry, sir; 'Royal Sovereigns' very hungry to-day; very good appetites they had, sir; wonder they left even the bones."

"You're laughing at me, you rascal; bring another chicken!"

"No more chickens, sir; roast lamb, would the lady like? hot or cold; green peas, new potatoes?" . .

"Your apple tart, sir. Ale, sir. Claret, ma'am." . .

Dinner disposed of, the coach is ready, but one of our passengers is missing. Has any one seen him? He went off, it seems, to see the cathedral, instead of having dinner. Fortunately for himself he comes hurrying up just as we are starting, and the guard hauls him up to his outside place by main force

"Tip us a tune," says the coachman to the guard, who, rendered sentimental by the steak and the bottle of stout he had for dinner in the bar, in company with the buxom barmaid, responds with "Believe me, if all those Endearing Young Charms," as we pass the frowning portal of Bootham Bar and bump along the very rough street of Clifton, York's modern suburb.

This is a thirteen-and-a-half mile stage from York to Easingwold, but although long, it is an easy one for the horses, if the coachman does not demand pace of them, on account of the dead level of the road. He very wisely lets them take their own speed, only now and then shaking the reins when they seem inclined to slacken from their steady trot. It is a lonely stretch of country, treeless, flat, melancholy; and the appearance of Easingwold is welcomed. At the "Rose and Crown" the new team is put in, and off we go again, the ten miles to Thirsk. At Northallerton the horses are changed for a fresh team at the "Golden Lion," and the fat coachman, assisted down with almost as much trouble as he was hoisted up, resigns the ribbons into the hands of another.

The usual knot of sightseers of the little town are gathered about the inn to witness the one event of the day, the arrival of the London coach. Among them one perceives the coachman out of a place; a beggar out at elbows; three recruits with ribbons in their hats, not quite recovered from last night's drink, and stupidly wondering how the ribbons got there; the "coachman wot is to take the next stage"; several errand boys wasting their masters' time; and a horsey youth with small fortune but large expectations, who is the idler of the place—the local man about town. There is absolutely nothing else for the inhabitants of Northallerton to do for amusement but to watch the coaches, the post-chaises and the chariots as they pass along the one long and empty street.

Our box-seat passenger leaves us here. Although he has, all the way down, shown himself anxious to be intimate with the successive coachmen, and has paid pretty heavily for the privilege of occupying that seat of honour, it has been of

no sporting advantage to him, for he is only a
Cockney tradesman, who has never even driven
a trap, let alone four-in-hand. So when each
whip in turn asked him the questions, conventional among whips, "whether he had his
driving-gloves on, and would like to take the
ribbons for the next few miles," he evaded the
offer by "not being in form," or not knowing
the road, or something else equally annoying to
the coachman, who, in not having an amateur of
driving on the box, thereby missed the canonical
tip of anything from seven shillings to half a
sovereign which the handling of the reins for
twenty miles or so was worth to the ordinary
sportsman.

Our new coachman, on our starting from
Northallerton, keeps the seat beside him vacant
He says he has a passenger for it down the road.
Tom Layfield, for that is the name of our present
charioteer, works the "Wellington" up and down
between this and Newcastle on alternate days,
Ralph Soulsby being the coachman on the other
Tom Layfield is a very prim-looking, tall and
spare man, tutor in coachmanship to many gentlemen on these last fifty-five miles; and it does not
surprise some of us when, passing Great Smeaton,
we are hailed by a very "down the road" looking
young man, whose hat is cocked at a knowing
angle, and whose entire get-up, from the gigantic
mother-o'-pearl buttons on his light overcoat to
the big scarf-pin in the semblance of a galloping
coach and horses, proclaims "amateur coachman."

It is the young squire of Hornby Grange, on the right hand, we are told, who is anxious to graduate in coaching honours, and to be mentioned in the pages of the *Sporting Magazine* by Nimrod, in company with Sir St. Vincent Cotton, the Brackenburys, and other distinguished ornaments of the bench.

"'Afternoon, squire," says Layfield, as that young sportsman swings into the seat beside him, and they talk guardedly about anything and everything but coaches, until Layfield asks—as though it had just occurred to him—if he would not like to "put 'em along" for a few miles He accepts, and is just about to take the reins over when the voice of a hitherto silent gentleman is heard from behind.

"I earnestly protest, coachman," he says, "against your giving the reins into the hands of that young gentleman, and endangering our lives. I appeal to the other passengers to support me," he continues, glancing round. "We read in the papers every day of the many serious, and some fatal, accidents caused by control of the horses being given to unqualified persons If you are well advised, young gentleman, you will relinquish the reins into their proper keeping; and you, coachman, ought to know, and do know, that you would be liable to a fine of any amount from £5 to £10, at the discretion of a magistrate, for allowing an unauthorised person to drive."

The coachman takes back the reins, and sulkily says he didn't know he had an informer up, to which the gentleman rejoins by saying that, so long as the coachman drives and performs the duty for which he is paid by his proprietors, he himself is not concerned to teach him proper respect; but he cannot refrain from pointing out, to the coachman in especial, and to the passengers generally, that it would have been the policy of an informer to allow the illegal act to be committed and then to lay an information. He was really protecting the coachman as well as the passengers, because it was well known that the road swarmed with informers, and continued infractions of the law could not always hope to go unpunished.

Every one murmurs approval, except the coachman and his friend, and the guard. The guard, as an official, is silent; the amateur coachman has a hot flush upon his face. The coachman, however, clearly sees himself to be in the wrong, and awkwardly apologises. Still, we all feel somewhat constrained, and, passing Croft Spa and coming to Darlington, experience an ungrateful relief when the champion of our necks and limbs leaves us there.

He is no sooner gone than tongues are wagging about him "Who is he? What is he? Do you know him?"

"Talks like a Hact o' Parlymint," says the coachman to his friend

"And a very good reason, too," says a man with knowledge. "he is a Justice of the Peace and Chairman of the Bench of Magistrates at

Stockton, which holds a higher jurisdiction than your bench, coachman. I think you've had a very narrow escape of parting with £10 and costs."

The guard has a few parcels to take out of the boot at the "King's Head," and a few new ones to put in, and then we're off for Rushyford Bridge, where the coach takes tea, and where we leave the amateur coachee at the "Wheatsheaf."

Durham and the coal country open out on leaving secluded Rushyford. Durham Cathedral, although itself standing on a height, has the appearance of being in a profound hollow as the coach, with the skid on, slowly creaks and groans down the long hill into the city. Changing at the "Three Tuns," the new team toils painfully up the atrociously steep streets to Framwellgate Bridge, where the river Wear and the stern grandeur of the Norman Cathedral, with the bold rocks and soft woods around it, blend under the westering sun-rays of a July evening into a lovely mellowed picture.

Chester-le-Street and Gateshead are ill exchanges for the picturesqueness of Durham, but they serve to bring us nearer our journey's end, and, truth to tell, we are very weary; so that, coming down the breakneck streets of Gateshead in the gathering darkness to the coaly Tyne and dear dirty Newcastle, with the hum of its great population and the hooting of its steamers in our ears, we are filled with a great content. " Give 'em a tune," says the coachman; and, the guard sounding a fanfare, we are quickly over the old town bridge, along the Side, and at the Turf Hotel, Collingwood Street. It is nearly ten o'clock. The journey is done.

Let us tot up the expenses per head :—

	£	s	d.
One outside place	3	10	0
Supper at Arrington Bridge .	0	2	6
Brandy and sandwiches at Huntingdon	0	3	0
Coachman, Huntingdon	0	1	6
,, Newark . . .	0	2	0
Breakfast, Doncaster	0	2	3
Guard, Doncaster .	0	2	6
Coachman, Ferrybridge . .	0	2	0
Dinner, York . .	0	3	6
Coachman, Northallerton .	0	1	6
Tea, Rushyford Bridge .	0	2	0
Coachman, Newcastle	0	2	0
Guard, Newcastle . . .	0	2	6
Total . . £4	17	3

CHAPTER IV

ONE of the greatest objections urged by the coaching interest against railways was their danger, and the certain loss of life on them in case of accident It was unfortunate that the opening of the Liverpool and Manchester Railway was the occasion of a fatal mischance that lent emphasis to the dolorous prophecies of coach-proprietors and the road interests in general; for on that day (September 15th, 1830) Mr. Huskisson, a prominent man in the politics of that time, met his death by being run over by the first train. It seems to ourselves incredible, but it was the fact, that there were those who ascribed this fatality to the wrath of God against mechanical methods of travelling. Then first arose that favourite saying among coachmen, "In a coach accident, there you are; in a railway accident, where are you?" The impression thus intended to be conveyed was that a coaching disaster was a very trifling affair compared with a railway accident. But was it? Let us see.

The Rev. William Milton, who in 1810 published a work on coach-building, lamented the great number of accidents in his time, and said that not a tenth part of them was ever recorded in the newspapers. He darkly added that the coach-proprietors could probably explain the reason. However that may be, the following pages contain a selection of the most tragical happenings in this sort, culled from the newspapers of the past. It does by no means pretend to completeness; for to essay a task of that kind would be to embark upon a very extensive work, as well as a very severe indictment of the coaching age. Moreover, it may shrewdly be suspected that many drowsy folk fell off the box-seats in the darkness, and quietly and unostentatiously broke their necks, without the least notice being publicly taken. Mere upsets and injuries to passengers and coachmen are not instanced here. Only a selection from the fatal accidents has been made.

1807.—Brighton and Portsmouth coach upset; coachman killed.

1810.—Rival Brighton and Worthing coaches racing; one upset; coachman killed.

1819.—"Coburg" (Brighton coach) upset at Cuckfield, on the up journey. The horses were fresh, and, dashing away, came into collision with a waggon. All the eleven outsides were injured. A Mr. Blake died next day at the "King's Head," Cuckfield, where the injured had been taken.

1826. *April.*—The Leeds and Wakefield "True Blue," going down Belle Hill with horses galloping, on the wrong side of the road, came into collision with a coal-cart. The coachman's skull was fractured, and he died instantly. One outside passenger's leg had to be amputated, and he died the next day. The recovery of another passenger was regarded as doubtful.

One of the more serious among coach accidents was that which befell the London and Dorking stage, in April 1826. It was one of those coaches that did not carry a guard. It left the "Elephant and Castle" at nine o'clock in the morning, full inside and out, and arrived safely at Ewell, where Joseph Walker, who was both coachman and proprietor, alighted for the purpose of getting a parcel from the hind boot. He gave the reins to a boy who sat on the box, and all would have been well had it not been for the thoughtless act of the boy himself, who cracked the whip, and set the horses off at full speed. They dashed down the awkwardly curving road by the church and into a line of wooden pailings, which were torn down for a length of twelve yards. Coming then to some immovable obstacle, the coach was violently upset, and the whole of the passengers hurled from the roof. All were seriously injured, and one was killed. This unfortunate person was a woman, who fell upon some spiked iron railings, "which," says the contemporary account, "entered her breast and neck. She was dreadfully mutilated, none of her features being distinguishable. She lingered until the following day, when she expired in the greatest agony." The gravestone of this unfortunate person is still to be seen in the leafy churchyard of Ewell, inscribed

A MIDNIGHT DISASTER ON A CROSS ROAD : FIVE MILES TO THE NEAREST VILLAGE.

After C. B. Newhouse.

to the memory of "Catherine, wife of James Bailey, who, in consequence of the overturning of the Dorking Coach, April 1826, met with her death in the 22nd year of her age."

1827 *December.*—The up Salisbury coach was driven, in the fog prevailing at the time, into a pond called the "King's Water," at East Bedfont, on Hounslow Heath. An outside passenger, a Mr. Lockhart Wainwright, of the Light Dragoons, was killed on the spot, by falling in the water. The pond was only two feet deep, but it had a further depth of two feet of mud, and it was thought that the unfortunate passenger was smothered in it. The four women inside the coach had a narrow escape of being drowned, but were rescued, and the coach righted, by a crowd of about a hundred persons, chiefly soldiers from the neighbouring barracks, who had assembled.

1832 *February 19th.*—Mr. Fleet, coachman and part-proprietor of the Brighton and Tunbridge Wells coach, killed by the overturn of his conveyance

1832. *October 30th.*—Brighton Mail upset at Reigate. Coachman killed on the spot. The three outsides suffered fractured ribs and minor injuries

In 1833 the Marquis of Worcester, a shining light of the road in those days, began that connection with the Brighton Road which afterwards produced the "Duke of Beaufort" coach, made famous by the coloured prints after Lambert and Shayer. He was passionately fond of driving, and was so very often allowed by the complaisant professional coachmen to "take the ribbons" that he at last fell into the habit of taking them almost as a matter of right. Of course, the jarveys who had relinquished the reins to him were always well remembered for their so doing; but there were those to whom money was not everything, and in whose minds the sporting instinct was less developed than a wholesome and ever-present fear of the penalties to which coachmen were liable if they permitted other persons to drive. There could have been no objection on the score of coachmanship, for the Marquis was an able whip; but the fact remained that he could not get the reins when he wanted them, and so in revenge set up two coaches on the Brighton Road, in alliance with a Jew named Israel Alexander. A paltry fellow, this Marquis, afterwards seventh Duke of Beaufort, to enter into competition with professional coachmen in order to satisfy a childish spite; not, at any rate, the high-souled sportsman that toadies would have one believe.

The coaches put on the road by this alliance were the "Wonder" and the "Quicksilver," both with intent to run Goodman, the proprietor of the "Times" coaches, off the route. The coachmen who tooled these new conveyances were, of course, always to give up the reins when my lord thought proper to drive, and so the revenge was complete. But the "Quicksilver," a fast coach timed to do the 52 miles in $4\frac{3}{4}$ hours, had not been long on the road before it met with

THE "BEAUFORT" BRIGHTON COACH.

After W J Sawyer.

a very serious accident, being overturned when leaving Brighton on the evening of July 15th. A booking-clerk, one John Snow, the son of a coachman, and himself a sucking Jehu, was driving, and upset the coach by the New Steyne, with the result that the passengers were thrown into the gardens of the Steyne, or hung upon the spikes of the railings in very painful and ridiculous postures. Goodman had the satisfaction of presently learning that the bad-blooded sportsman and his partner lost some very heavy sums in compensation awards.

The "Quicksilver" was thereupon repainted and renamed, and, under the alias of the "Criterion," resumed its journeys. But ill-fortune clung to that coach, for on June 7th, 1834, as it was leaving London, it came into collision with a brewer's dray opposite St. Saviour's Church, Southwark. A little way on, down the Borough High Street, the coachman was obliged to suddenly pull up the horses to avoid running over a gentleman on horseback, whose horse had bolted into the middle of the road. The sudden strain on the pole, already, it seems, splintered in the affair with the dray, broke it off. It fell, and became entangled with the legs of the wheelers, who became so restive and infuriated that attempts were made to put on the skid; but before that could be done the coach overturned. Sir William Cosway, who was one of the outsides, and was at that moment attempting to climb down, was pitched off so violently that his skull was fractured, so that he died in less than two hours afterwards. A Mr. Todhunter "sustained" (as the reporters have it) a broken thigh.

1834.—The London and Halifax Mail came into collision with a bridge, five miles from Sheffield The coachman, Thomas Roberts, was killed.

The Wolverhampton and Worcester coach, in avoiding a cart coming down a hill near Stourbridge, was upset, and a passenger killed.

October —A wheel came off one of Wheatley's Greenwich coaches at London Bridge, and one gentleman was killed.

1835 *August* —The Liverpool "Albion" fell over on entering Whitchurch, through a worn-out linchpin. A lady inside passenger was disfigured for life.

June.—The Nottingham "Rapid" upset, three miles from Northampton, through the breaking of an axle. A girl's leg crushed, and afterwards amputated.

November —The Newcastle and Carlisle Mail upset, two miles from Hexham. Aiken, the coachman, killed.

December 25th.—The down Exeter Mail upset on Christmas night, on nearing Andover, through running against a bank in the prevailing fog. Austin, the coachman, killed.

1836. *June.*—The up Louth Mail nearly upset by stones maliciously placed in the road by some unknown person, near Linger House bar Rhodes, the guard, was thrown off and seriously injured.

In September, 1836, a shocking accident befel the down Manchester " Peveril of the Peak," five miles from Bedford. The coach turned over, and a gentleman named O'Brien was killed on the spot. The coachman lay two hours under the coach, and died from his injuries.

The next disaster on our list was caused by a drunken coachman's dazed state of mind. Early on a Sunday morning in June, 1837, the Lincoln and London Mails met and came into collision at Lower Codicote, near Biggleswade. The driver of the up mail, Thomas Crouch, was in a state of partial intoxication at the time, and owing to a curve in the road, and the wandering state of his faculties, he did not observe the approach of the other mail The result was that, although the coachman of the other made room for him to pass, the two coaches came into violent collision. The coach driven by Crouch was turned completely round, ran twenty or thirty yards in a direction opposite to that it was originally taking, and finally settled in a leaning posture in the ditch. Crouch was so injured that he died a few hours afterwards. The passengers were not much hurt, but two horses were killed.

On September 8th, a coachman named Burnett was killed at Speenhamland, on the Bath Road. He was driving one of the New Company's London and Bristol stages, and alighted at the " Hare and Hounds," very foolishly leaving the horses unattended, with the reins on their backs He had been a coachman for twenty years, but experience had not been sufficient to prevent him thus breaking one of the first rules of the profession. He had no sooner entered the inn than the rival Old Company's coach came down the road. Whether the other coachman gave the horses a touch with his whip as he passed, or if they started on their own accord, is not known, but they did start, and Burnett, rushing out to stop them, was thrown down and trampled on so that he died.

Of another kind was the fatal accident that closed the year on the Glasgow Road. On the night of December 18th, the up Glasgow Mail ran over a man, supposed to have been a drunken carter, who was lying in the middle of the highway.

1837. *August.*—The up Glasgow Mail, the up Edinburgh Mail, the Edinburgh and Dumfries, and the Edinburgh and Portpatrick Mails all upset the same night, at different places.

1838. *August.*—The London to Lincoln Express met a waggon at night, at Mere Hall, six miles from Lincoln. The coachman called to the waggoner to make room, and a young man who, it is supposed, was asleep on the top, started up, and rolled off. The waggon-wheels went over and killed him

September.—The Edinburgh and Perth "Coburg" was the subject of a singular accident. Passengers and luggage were being received at Newhall's Pier, South Queensferry, when the leader suddenly turned round, and before the coachman and guard, who were stowing luggage, could render assistance, coach and horses disappeared over the quay-wall. Some of the outsides saved their lives by throwing themselves on the pier, but the four insides were less fortunate. Two of them thrust their heads through the windows, and so kept above the sea-water; the other two—a Miss Luff and her servant—were drowned. One outside, who had been flung far out into the sea, could fortunately swim, and so came ashore safe, but exhausted. Nine years later, February 16th, 1847, a similar accident happened to the Torrington and Bideford omnibus, when the horses took fright and plunged with the vehicle into the river from Bideford Quay. Of the twelve passengers, ten were drowned.

October.—The " Light Salisbury," having met the train at Winchfield Station, proceeded to Hurstbourne Hill, between Basingstoke and Andover, where the bit of one of the horses caught in the pole and the coach was immediately overturned. One passenger died the same afternoon, and another was taken to his house at Andover without the slightest hope of recovery. A young woman's leg was broken, and two other passengers' limbs were smashed.

The railway journals, which had even thus early sprung into flourishing existence, did not fail to notice the increasing number of coaching accidents, the *Railway Times* with great gusto reporting twenty in a few weeks. The prevalence of these disasters was a cynical commentary upon the " Patent Safety " coaches running on every road, warranted never to overturn and doing so with wonderful regularity, and on those coaching prints noticed by Charles Dickens—" coloured prints of coaches starting, arriving, changing horses ; coaches in the sunshine, coaches in the wind, coaches in the mist and rain, coaches in all circumstances compatible with their triumph and victory ; but never in the act of breaking down, or overturning."

The last years of coaching were, in fact, even more fruitful in accidents than the old days. Especially pathetic were the circumstances attendant upon the disaster that overtook the " Lark " Leicester and Nottingham Stage on May 23rd, 1840. The coach was on its last journey when it occurred, for the morrow was to witness the opening of the railway between those places. Like most of these last trips, the occasion was marked by much circumstance. Crowds assembled to witness the old order of things visibly pass away, and Frisby, the coachman, had dolefully tied black ribbons round his whipstock, to mark the solemnity of the event. Unfortunately, that badge of mourning proved in a little while to be only too appropriate, for the well-loaded coach had only gone about a mile and a half beyond Loughborough when Frisby, who had been driving recklessly all the way, and had several times been remonstrated with, overturned it at Coates' Mill. A Mr. Pearson and another were killed. Pearson, who had especially come to take part in this last drive, was connected with the " Times " London and Nottingham coach. He had been seated beside

A QUEER PIECE OF GROUND IN A FOG: "IF WE GET OVER THE RAILS, WE SHALL BE IN A UGLY FIX."

After C. B. Newhouse.

Frisby, and had several times warned him, without avail. His thighs were broken, and he received a severe concussion of the brain, from which he died at midnight. Frisby himself was crippled for life

The pitcher goes oft to the well, but at last it is broken; and so likewise the coachmen who, winter and summer, storm or shine, had driven for almost a generation over the same well-known routes, at length met their death on them in some unforeseen manner. A striking instance of this was the sad end of William Upfold—" unlucky Upfold "—who was coachman of the "Times" Brighton and Southampton Stage, a coach which ran by way of Worthing and Chichester. He was a steady and reliable man, fifty-four years of age, and had been a coachman for thirty-five years, when fatal mischance slew him on a February night, 1840. A singularly long series of more or less serious accidents had constantly attended him from 1831. In that year his leg was broken in an upset, and he had only just recovered and resumed his place when the coach was overturned again, this time through the breaking of an axle. The injuries he received kept him a long time idle. Again, in January 1832, at Bosham, the furies were eager for his destruction. He got off at the wayside inn, and left the reins in the hands of a passenger, who very foolishly alighted also, a minute or so later. When Upfold saw him enter the inn he hastily left it; but the horses had already started,

In trying to stop them he was kicked on the leg, and fell under the wheels, which passed over him and broke the other leg.

Poor Upfold recovered at last, and might have looked forward to immunity from any more accidents; but Fate had not yet done with him. When nearing Salvington Corner, one night in February 1810, he was observed by Pascoe, a coachman who was with him, to pull the wrong rein in turning one of the awkward angles that mark this stretch of road.

"Upfold, what are you at with the horses?" he asked.

"I have pulled the wrong rein," said Upfold.

"Then mind and pull the right one this time," rejoined Pascoe; but scarcely had he said it when the coach toppled over. Nearly every one was hurt, but Upfold was killed. His pulling the wrong rein was inexplicable. The unfortunate man knew the road intimately, and the witnesses declared he was absolutely sober; and so the country-folk, who knew his history and how often accidents had come his way, were reduced to the fatalistic remark that "it had to be."

1841 *November 8th.*—Rival coaches leaving Skipton started racing on the Colne and Burnley road. The horses of one grew unmanageable and ran away. The passengers, alarmed, began to jump off, and a Manchester man, name unknown, who had been sitting beside the coachman, laid hold of the reins to help the coachman pull the horses in. In doing so, he pulled their heads to one side, and they dashed with appalling force into a blank wall. He was killed on the spot. All the passengers who had jumped off were more or less seriously injured ; but a woman and a boy, who had remained quietly in their seats on the roof, were unhurt.

1842. *January* 17*th*.—The "Nettle," Welshpool and Liverpool coach, overturned by a stone near Newtown. Mr. Jones, of Gorward, Denbighshire, a Dissenting minister, going to live at Kerry, Montgomeryshire, was thrown off the roof. He died two days later of his injuries, in great agony.

December 28*th*.—The Mail, coming south from Caithness-shire, broke an axle at Latheronwheel Bridge, and Donald Ross, the coachman, was dashed from his box over the bridge into the rocky burn, thirty feet below, and killed. The guard had a narrow escape. Fortunately, there were no passengers.

1843. *February* 18*th*.—The Cheltenham and Aberystwith Mail left the "Green Dragon," at Hereford, on its way, and proceeded as usual to St. Owen's turnpike-gate. The gate was open, as a matter of course, for the Mail, but the boisterous wind blowing at the time sent it swinging back across the road as the Mail passed. It hit the near wheeler a violent blow and broke the trace and the reins. Then rebounding, it struck the body of the coach with such force that Eyles, the coachman, was thrown off the box and killed. The horses, thoroughly terrified, then ran away, and, meeting some donkey-carts on the road, ran into them, injuring some old women driving from market One of them subsequently died from her hurts.

March 22*nd.*—The Norwich Day Coach upset at Brentwood. The coachman, James Draing, who was also proprietor, was killed.

April 21*st.*—The Southampton and Exeter Mail upset in the New Forest, two miles from Stony Cross, by the horses, frightened at an over-turned waggon, running the coach up a bank. Cherry, the coachman, met a dreadful death, his head being literally split in two A sub-scription of £350 was raised for his widow and six children.

May 1*st.*—The "Red Rover," Ironbridge and Wolverhampton coach, upset half a mile from Madeley. One passenger, name unknown, killed. He was described as "a very stout gentleman, apparently about sixty years of age, dressed in an invisible green coat and great-coat of the same colour."

June 26*th.*—William Cooke, guard of the Worcester coach, fell off his seat and was killed.

September 16*th* —The Ludlow and Bewdley "Red Rover" overturned by the breaking of the front axle. The coach was going slowly down-hill at the time, and the wheel had the slipper on. It was a heavily-loaded coach, and all the outsides were violently thrown. A Mr. Thomas, a native of Ludlow, fifty-seven years of age, retired from business, was so seriously injured that he died

ROAD VERSUS RAIL.

After C. Cooper Henderson 1845.

next day. At the inquest a deodand of £30 was placed on the coach.

From this time forward the records of coaching accidents grow fewer, and occur at longer intervals, but only because coaches themselves were being swiftly replaced by the railways, which had by now come largely into their kingdom. Railway accidents took their place, and the coaching artists began to paint, and the printsellers to publish, pictures like that of "Road *versus* Rail" engraved here, showing a very smart and well-appointed coach bowling safely along the road, while a railway accident in progress in the middle distance attracts the elegant and rather smug attention of coachman and passengers

Every one now forgot the numerous casualties of the old order of things—save, indeed, the bereaved and the maimed, suffering from the happenings of pure mischance, or from the drunken or sporting folly of the coachmen.

But to the very last, in those outlying districts to which the rail came late, and where the coaches continued to ply regularly until the 'fifties, the tragical possibilities of the road were insistent, confounding the thorough-going sentimentalists to whom the old times were everything that was good, and the new, by consequence, altogether bad. Listen to the moving tale of the Cheltenham and Aberystwith down mail on a wild night "about" 1852, according to the vague recollection for dates of Moses James Nobbs.

Although torrents of rain had been falling

and the night was pitch dark, all went well with
the mail until nearing the Lugg Bridge, near
Hereford, where the little river Lugg, rushing
furiously in spate to join the Wye, had under-
mined the masonry. No sooner did the horses
place their weight upon it than the arch gave
way, and the coachman, Couldery the guard, and
the one passenger, were precipitated into the
torrent and swept away for more than a mile
down stream. It was midnight when the accident
happened, and until daybreak the three, at
separate points, clung to rocks and branches, from
which they were then rescued by search-parties.
The coachman and guard recovered from the
exposure, but the passenger died.

Charles Ward, that fine old coachman, who
kept on the road in Cornwall for many years
after coaching had ceased over the rest of England, tells amusingly of the happening that befell
the cross-country Bath and Devonport Mail, in
some year unspecified. It might have been a
most serious accident, but fortunately ended
happily. The coach was due to arrive at Devonport at eleven o'clock at night. On this particular occasion all the outside passengers, except
a Mrs. Cox, an "immense woman," who kept
a fish-stall in Devonport Market, had been set
down at Yealmpton, where the coachman and
guard usually had their last dram. They went,
as usual, into the inn, and very considerately
sent out to Mrs. Cox a glass of "something
warm," it being a very cold night. The servant-

girl who took out that cheering glass was not able to reach up to the roof, and so the ostler, who was holding the horses' heads, very imprudently left them, to do the polite, when the animals, hearing some one getting on the coach, and thinking (for coach-horses did actually do something like it) that it was the coachman, started off, and trotted at their ordinary speed the whole seven miles to the door of the "King's Arms" at Plymouth, where they were in the habit of stopping to discharge some of the coach-freight. On their way they had to cross the Laira Bridge and through the toll-bar, and did so, keeping clear of everything on the road in as workmanlike a manner as though the skilfullest of whips was directing their course. Mrs. Cox, however, was terrified. Afraid to scream lest she should startle the horses, she had to content herself with gesticulating and trying to attract the attention of the people met or passed on the road. When the horses drew up in an orderly fashion at the "King's Arms," and the ostlers came bustling out to attend to their duties, they were astonished to see no one but the affrighted Mrs. Cox on the outside, and two inside passengers, who had been in total ignorance of what was happening. The coachman and guard, in a very alarmed state, soon came up in a post-chaise. It took many quarterns of gin to steady the nerves of the proprietress of the fish-stall, and the incident became the chief landmark of her career.

We will conclude this chapter of accidents on this lighter and less sombre note, and tell how humour sometimes remained in the foreground even if the possibilities of tragedy lurked threatening in the rear. The tale used often to be told on the Exeter Road how, on one occasion, when Davis was driving the up "Quicksilver" Mail between Bagshot and Staines on a dark night, he ran into some obstruction, and the coach was upset into the adjoining field, fortunately a wet meadow. The "insides" were asleep at the time, and they naturally awoke in the wildest alarm. One, who did not grasp the situation, called out, "Coachman, coachman, where are we?" "By God, sir," replied Davis, "I don't know, for I was never here before in all my life!" Happily, nobody and nothing was hurt, and in twenty minutes the coach was away, making up for lost time.

CHAPTER V

A GREAT CARRYING FIRM. THE STORY OF
PICKFORD AND CO.

To the incurious public, who are as familiar with
the name of " Pickford's " as with that of their
favourite morning newspaper, and to whom the
sight of one of Pickford's vans is a mere commonplace of daily life, this great carrying firm is just
a part of our modern commercial system To
suggest to that favourite abstraction—the "average
man "—so commonly cited, that Pickford's is a
firm whose origin is to be traced back two hundred
and fifty or three hundred years would be a rash
thing. He would tell you that this is a firm of
railway carriers, and that, as railways are not yet
a hundred years old, Pickford's certainly cannot
be two centuries older.

Thus do later changes overlie and conceal
earlier methods of business.

Our average citizen would be wrong in two
things · in his premisses, that the firm is wholly
one of railway carriers; and in his conclusions,
that it came into existence with railways themselves. The origin of Pickford's is, indeed, lost in
the mists that gather round the social and commercial life of the early seventeenth century; for

the beginnings of the business go back to that time when the original firm of packhorse carriers was established, to whose trade the Pickfords succeeded, by purchase or otherwise, about 1730 Traditions only survive of those long-absorbed carriers, whose packhorse trains originally plied on the hilly tracks between Derby and Manchester " about two hundred and fifty or three hundred years ago," as we vaguely learn No documentary or other evidence exists on which to found an account of them. What would we not give to be able to recover from the romantic past the story of those old-time carriers, contemporary with the famous Hobson himself, beyond comparison the most celebrated of all these old men of the road !

But all records have been destroyed. When the several changes were made that from time to time altered the constitution of the business, the papers and documents relating to past transactions were cast aside as waste-paper, and there was none among the people of those times who thought it worth while, for the interest and instruction of posterity, to set down what he knew of the current history of the concern That this should have been the case is no matter for surprise. The past or the future interests many to whom the present is only something from which to escape, as commonplace and dull. That man who is not glad, when the business day is done, to leave for home and straightway dismiss all thoughts of his business from his mind is rare indeed ; and still more rare he who finds interest, beyond mere money-getting, in the daily commerce by which he lives and prospers.

About 1770 Matthew Pickford, the representative, in the second or third generation, of that family in this olden firm, is found established in Manchester, a town then making rapid industrial progress, and affording great scope for the carrying trade, already, for some years past, conducted by waggons , but we do not obtain any details of his business until November 16th, 1776, when he issued the following advertisement, afterwards inserted in *Prescott's Manchester Journal* for Saturday, January 4th, 1777 :—

"THIS is to acquaint all Gentlemen, Tradesmen, and Others, that Mat. Pickford's Flying Waggons to London in Four Days and a Half

Set out from the Swan and Saracen's Head, in Market Street Lane, Manchester, every Wednesday, at Six o'clock in the Evening, and arrive at the Swan Inn, Lad Lane, London, the Tuesday noon following ; also set out every Saturday at the same Hour, and arrive there on Friday noon following. Set out from London every Wednesday and Saturday, and arrive at Manchester every Tuesday and Friday ; which carry goods and passengers to and from Manchester, Stockport, Macclesfield, Leek, Blackburn, Bolton, Bury, Oldham, Rochdale, Ashton-under-Line, and places adjacent.

" N.B.—M Pickford will not be accountable for any Money, Plate, Watches, Jewels, Writings, Glass, China, etc , unless entered as such, and paid for accordingly.

" Constant attention at the above Inns in London and Manchester, to take in Goods, etc."

It will be noticed that these " four days and a half" trips, although performed by " Flying Waggons," and presumably much swifter than some earlier ones of which we have no record, were only four and a half days in a very special sense, and by the exercise of some peculiar method of reckoning whose secret has not descended to us. It might seem, to the person of ordinary intelligence, that these were really itineraries of rather more than five days and a half; but the Sunday was doubtless a day of rest for the waggoners, as for most others in those times.

In 1780, according to the evidence afforded by an old billhead, still preserved, Matthew Pickford was carrying on business in conjunction with Thomas, his brother, and in this partnership they continued to trade for many years.

Meanwhile, the manufacturing industries of Lancashire and the north-west had grown enormously, and canals were already being dug to aid the transport of goods. We have no means of knowing in how far the Pickfords took advantage of the early canals in the Midlands, but that they availed themselves very greatly of the opportunities afforded by them of extending their business seems unlikely, in view of the position in 1817, when they admitted Joseph Baxendale as a partner into the concern.

Joseph Baxendale was thirty-two years of age when he became partner in the firm of Pickford & Co. He was born in 1785, the son of Josiah Baxendale, of Lancaster, and had already seen something of business as partner in the concern of Swainson & Co., calico-printers at Preston, whose firm he left to seek those wider activities for which his active mind longed. For there was something adventurous in his blood, which would by no means permit him merely to take the sedentary part of a capitalist in any enterprise in whose fortunes he might acquire a share. An opportunity thoroughly suited to his temperament was this which offered, of becoming a partner in the already old-established firm of Pickford's. We have now no means of knowing precisely on what terms he joined the two brothers, but whatever the pecuniary consideration may have been, enough survives to tell us that his youthful activities and his keen business intelligence were prominent in what he brought into the firm. For many years Matthew and Thomas had borne the whole conduct of the business, and it was now desirable, both by reason of their advancing years and the natural growth of the commercial activities of the country, that they should have, allied with them, one who, alike by inclination and urged by business interests, would scour the country, supervising and organising, as they no longer found it possible to do.

Baxendale found plenty of work of this nature awaiting him. The staff of horses which the Pickfords had found sufficient for their needs in bygone years had been little, if at all, increased, although a period of great trade-expansion had set in; and a total lack of efficient supervision over agents and carmen had resulted in the carrying business being dilatory and untrustworthy. Under these circumstances, it is not surprising that rival firms had begun to threaten the very existence of Pickford's, declining under the nerveless rule, by which the needs of the time were not understood

It was soon impressed upon the new partner's active and penetrating intelligence that the requirements of the time, and still more the requirements of the succeeding years, imperatively demanded a thorough reorganisation — more thorough, perhaps, than the old partners were altogether ready to concede. He soon acquired entire control, and the Pickfords, unable or unwilling to meet new times with new methods, left their already historic business and its destinies in his hands.

He speedily altered the aspect of affairs Soon he had close upon a thousand horses, all his own, on the great roads between London and the north-west; while advertisements were issued, announcing "Caravans on Springs and Guarded, carrying Goods only, every afternoon at 6 o'clock," from

London and Manchester, taking only 36 hours
to perform the 186 miles

To this, then, the " caravan " had come at last.
Travellers from the Far East had originally
brought the word to England. They had seen
the Persian *kārwāns* toiling under those torrid
skies—covered waggons in whose shady interiors
the poor folks travelled; and when the first stage-
waggons were established in England, they were
often known by an English version of that name.
Some of the caravans of the late seventeenth
century were, however, by no means the rough-
and-ready affairs generally supposed, if we may
judge from the description of one offered for sale
in the *London Gazette* of May 6th, 1689. This,
according to the vendors, was :—

" A Fair easie going Caravan, with a very
handsom Roof Brass Work, good Seats. Glasses
on the sides to draw up; that will carry 18 Persons, with great Conveniency for Carriage of
Goods, so well built that it is fit for Carriage
of all manner of Goods—to be sold."

But there was one more change before the
caravan in 1817. Already the popular voice,
unwilling to enunciate three syllables when one
could be made to serve, had clipped the name
to "van," and as vans all covered vehicles of the
kind have been known ever since.

At the time when Baxendale appeared upon
the scene the headquarters of the business were
still at Manchester, and the London establishments had been for many years past at the

"Castle," Wood Street, and the "White Bear," Basinghall Street. To the first house, then a a galleried inn of the ancient type, at the corner of Wood Street and what is now Gresham Street, but was then Lad Lane, the London and Manchester waggons and caravans resorted; and to and from the "White Bear" went the Leicester and Nottingham traffic.

Coming with a fresh mind to the carrying problems that confronted the firm, the new partner decided that London, and not Manchester, ought to be its central point, and so soon as he obtained control he accordingly removed the head offices to the Metropolis. Canal-traffic, too, engaged his earnest attention, and the scope of the firm's activities were extended enormously in that direction. The Regent's Canal was opened in 1820, and when that opening took place the newly built wharves of Pickford & Co. were ready, beside the City Basin. To and from that point came and went the water-borne trade, in the fly-boats of the firm, simultaneously with the fly-vans on the roads.

These developments brought other changes, and in 1826 the existing headquarter offices of Pickford & Co. were built in Gresham Street, adjoining the "Castle" Inn.

It will be interesting to see what was the cost of carriage of goods at this period. It was the carriers' Golden Age, when, for distances of a little over a hundred miles from London—as, for example, Leicester and Birmingham—the

JOSEPH BAXENDALE.

From the portrait by E. H. Pickersgill, R.A.

carriage of goods by waggon or caravan could be charged at 5s. per cwt., or £5 per ton; when by coach the rates for small parcels were 1d. a pound; and even by canal—that last effort in cheap transport before railways—the charges were 2s. 9d. per cwt., or £2 15s. per ton.

He who reorganised the old business of Pickford's demands extended notice in these pages. A portrait of him, a three-quarter length, painted by Pickersgill, R A., about 1847, has the illusion common to all three-quarter-length portraits of giving an appearance of great stature. Mr. Baxendale was a man of broad shoulder, and not above the middle height. While in many respects a good portrait of him, it is said by those who knew him best to fail in not giving expression to the native kindliness and humour that underlaid his keen business instincts. "Cheerful and witty in conversation," says one who knew him well, " he ever had a word of encouragement for the youngsters, and was universally beloved by those whom he employed."

To those who served him to the best of their ability he was a never-failing friend, and, at a time when business firms did not usually trouble themselves about the comfort of their servants, took pains to secure their well-being. In the galleries of the old "Castle" Inn he constructed a coffee- and club-room for his carmen, and provided similar conveniences at his other establishments. The old inn has long been demolished, but the headquarters of the firm still remains next

door, and adjoins the modern Railway Goods
Receiving Office of the "Swan with Two Necks,"
built on the site of the old coaching establishment
of Chaplin's.

Never was such a man for improving maxims
as Joseph Baxendale. He was a great admirer
of *Poor Richard's Almanack* and its racy maxims,
written by Daniel Webster, and carefully caused
a broadsheet containing a selection of them to be
printed. He also tried his own hand at composing
pithy sentences on the virtues of punctuality and
method, and caused leaflets of these, together
with *Poor Richard's* homely literature, to be
circulated and posted in all conspicuous places
in the establishments of Pickford & Co. in London
and the provinces, and on the roads and canals
where his vans travelled or his fly-boats voyaged.
Here is one of his compositions in this way :—

| TIME LOST
CANNOT
BE
REGAINED | THE
IMPORTANCE
OF
PUNCTUALITY | NEVER
DESPAIR
——
NOTHING
WITHOUT
LABOUR |

METHOD is the very Hinge of Business; and
there is no Method without Punctuality.
Punctuality is important, because it subserves the
Peace and good Temper of a Family : The want
of it not only infringes on necessary Duty, but
sometimes excludes this Duty. The Calmness of
Mind which it produces, is another Advantage

of Punctuality: A disorderly man is always in
a hurry; he has no time to speak to you, because
he is going elsewhere; and when he gets there,
he is too late for his business; or he must hurry
away to another before he can finish it. Punctuality gives weight to Character "Such a man
has made an Appointment:—then I know he
will keep it." And this generates punctuality in
you; for, like other Virtues, it propagates itself.
Servants and Children must be punctual, where
their Leader is so. Appointments, indeed, become
Debts. I owe you Punctuality, if I have made
an Appointment with you: and have no right to
throw away your time, if I do my own.

Of course, this good advice and insistence upon
its being followed would have been of little avail
had the author of it not been continually alert
to see that his instructions took root. *He*, at any
rate, practised what he preached, and rose early,
was diligent all day, and went late to bed. As a
business man whose business was conducted over
a large stretch of country—extending chiefly in a
diagonal line two hundred miles long, between
London and Liverpool—he knew that only by
personal supervision and by great and unwearied
exertions in travelling could his subordinates be
kept in a state of efficiency, and he accordingly
was always travelling. By post-chaise or by
private carriage he flew, day and night, along
the great roads between London and Holyhead,
and London, Derby, Manchester and Liverpool;

appearing, suddenly and unexpectedly, at some great town-warehouse of the firm, or some wayside office or place of call, and often springing, as it were, out of the void, to encourage some diligent servant, or (it is to be feared) more often to reprimand a lazy and inefficient one None could predicate his movements or where he might be at any given time ; save indeed those with whom he had made appointments, and they knew, after only a short acquaintance, that the sun was scarce more likely to rise and set according to the calendar than Joseph Baxendale was to redeem his promise of any such assignation.

Forsaking for awhile the roads and his establishments along them, he would next appear on the canals on whose sullen waters his fly-boats flew, and pay flying visits of inspection to the many wharves along their course. These water expeditions were made in a vessel especially constructed—a " canal-yacht " called the *Lark*, whether significantly named in allusion to the early-rising habits of its owner we do not know. It was this boat, according to the still surviving tradition, he lent to the Earl of Derby on an occasion when Lady Derby was in London, too ill to travel by road to Knowsley, where, according to the doctor's advice, she should be removed. In it she travelled all the way down to Lancashire, along the canals.

Another surviving tradition, and one that speaks well for the quality of the horses that drew the fly-boats—and perhaps even better for the keenness of the sporting instincts of the official concerned—tells how Mr. Baxendale, on coming to Braunston, a Northamptonshire village on the Grand Junction Canal, discovered that the man who should have been in charge of his wharf there had gone hunting, mounted on one of the firm's towing-path steeds Records of that time do not tell us of that sportsman's return, or of the reception that met him.

It was perhaps a consequence of the strenuous rule then obtaining that, at a time when the great roads to the north were blocked by the historic snowstorm of Christmas 1836, when the stage-coaches and the mails were buried in the drifts, Pickford's Manchester Flying Van was first through. We may suppose that the horses were better specimens than those pictured here, from an old painting, which represents the fly-van team as a very sorry one indeed, comparing badly with the sturdy animals who are seen drawing the van in the first picture.

It would be a mistake to think that Baxendale's ways with his staff were merely those of the strict disciplinarian, only anxious to obtain the utmost from them His kindliness was perhaps his strongest point, and Pickford's under his rule began the practice of recognising the loyalty and hard work of their servants by pensioning them on their retirement—a policy that still does honour to the firm.

Under this vigorous sway Pickford's grew and prospered, and by the time when railways first

loomed threatening upon the horizon of the
carriers' and coachmen's outlook, commanded the
bulk of the goods traffic between London and the
Midlands, alike by road and canal. That was a
period above all others when a clear head was
requisite. It appeared to many to be a choice
between giving up business or fighting the en-
croachments of steam. To the few, of whom
Baxendale was one, the issues were more varied
and hopeful. He foresaw that railways must
succeed, and that, since to fight them would be
hopeless, the best thing to do would be to work
with them as far as possible. The business
need not be injured; indeed, he saw that it must
needs share in whatever prosperity attended the
railways. Only methods must be changed. But
to reorganise a vast business only just, after
thirteen years of unwearied effort, re-established
on new and improved lines, must have seemed a
hard necessity. However, when the Liverpool
and Manchester Railway, the second line in the
country, was about to be opened, in 1830, he
perceived that although the road traffic must cease
between the two terminal points of a railway,
yet there must be some agency prepared to collect
goods, and deliver them to or convey them from
the railway stations. He saw, too, an inevitable
increase in the volume of traffic, and very pru-
dently resolved to obtain a share of it by throwing
in his lot with the railway people, who were
themselves not so assured of instant success as
to repel so unexpected an offer, and welcomed the

PICKFORD AND CO.'S ROYAL FLY-VAN, ABOUT 1820. *From a contemporary painting.*

proposed alliance. The same attitude was adopted
towards the Grand Junction Railway and the
London and Birmingham. In this far-seeing policy
Baxendale was at one with William Chaplin, who
at an early period in the history of railway enterprise had called upon him and asked him what
his views were on this vital question. Chaplin
withdrew his coaches when the London and
Birmingham Railway was opened, and Pickford's
fly-vans and fly-boats ceased to run. In return
for these really valuable services, Pickford's, and
Chaplin and his coaching ally, Horne, who had
been equally complaisant, acquired shares in the
town and country carrying agencies for what in
1815 became an amalgamation of railway interests
under the style and title of the "London and
North-Western Railway." Unused as these new
railway people were to the business of handling
goods, they were glad enough that Mr. Baxendale
should organise that class of traffic for them, and,
as we have already said, really welcomed the aid
thus somewhat unexpectedly forthcoming, although
outwardly adopting a self-sufficient and omnipotent
attitude. He became organising goods-manager,
and contributed the services of his staff to the
work, but resigned when everything had been
duly set going to devote himself to his own
business. He it was who drew up the documents
still used in the goods departments of railways to
this day, in all essentials unaltered.

Meanwhile his anticipations were justified by
the course of events. Railways did but alter the

methods of the carrying trade. They not only
did not destroy it, but, in the altered shape it took,
increased it fifty-fold. No fewer than twenty-one
district managers became necessary to the conduct
of the business, which at length gave employment
to between three and four thousand people.

The central figure of this successful reorganisation became, like William Chaplin, a power
in the railway world. He was for some years
Chairman of the South-Eastern Railway, and in
that capacity strongly urged the purchase of
Folkestone Harbour, an undertaking then in the
market. His co-directors did not at the time
agree with the proposal, but eventually came
round to his way of thinking, and brought up the
subject again. Meanwhile he had privately purchased the harbour. The high sense of duty that
characterised him led to his considering that, as
Chairman of the Railway Company, and as therefore trustee of the interests of the proprietors, he
could not retain the property, and he accordingly
transferred it at the price he had given. He
was at the same time a director of the Great
Northern Railway of France, but was in 1848,
in consequence of a severe illness, obliged to
resign some of these activities, together with the
detailed management of Pickford's, which he then
left in the hands of his three sons, but never gave
up control of the business. He had in the meantime purchased an estate at Woodside Park,
Whetstone, where he resided. He died there,
March 24th, 1872, in his eighty-seventh year.

The portrait of him, as he was in the full vigour of his manhood, hangs amid the old-time relics still cherished in the Gresham Street offices— among the muskets and the blunderbusses carried by the guards of his fly-vans in the old days of the road

CHAPTER VI

ROBBERY AND ADVENTURE

THE whole art and mystery of coach-robbing began to be studied at a very early date. In the *London Gazette* during 1684 we find the following extremely explicit advertisement :—

"A GENTLEMAN (paffing with others in the Northampton Stage Coach on Wednefday the 14th inftant, by Harding Common about two miles from Market-street) was fet upon by four Theeves, plain in habit but well-horfed, and there (amongft other things) robbed of a Watch ; the defcription of it thus, The Maker's Name was engraven on the Back plate in French, Gulimus Petit à Londres ; it was of a large round Figure, flat, Gold Enamelled without, with variety of Flowers of different colours, and within a Landskip, and by a fall the Enamel was a little cracked ; It had alfo a black Seale-Skin plain Cafe lined with Green Velvet. If any will produce it, and give notice to Mr. Samuel Gibs, Sadler near the George Inn Northampton, or to Mr. Crofs in Wood Street, London, he fhall have a Guinea reward."

It is to be feared that the gentleman who thus mourned his watch never regained it.

From this time forward, until well into the nineteenth century, highwaymen and the highway-robbery of postboys, stage-coaches, post-chaises, and all sorts and conditions of wayfarers became commonplaces of travel. Dick Turpin's name has acquired an undue prominence, on account of Harrison Ainsworth elevating him upon a pedestal, as the hero of a romance, but his was really neither a prominent nor an heroic figure Innumerable other practitioners surpassed him. Claude Du Vall, who robbed and danced on Hounslow Heath; Abershaw, the terror of the Surrey Commons; Captain Hind, soldier and gentleman, warring with authority; Boulter, whose depredations were conducted all over the kingdom, the "Golden Farmer" on the Exeter Road, outside Bagshot: all these and very many more were infinitely superior to Turpin, and, as they phrased it, "spoke to" the coaches with great success during their brief but crowded career. Nowadays, we hear much of overcrowded professions; but those of the Army, the Church, and the Law are by no means so crowded as were the ranks of the liberal profession of highway robbery in the brave nights of crape mask and horse-pistols at the cross-roads on the blasted heaths which then encompassed the Metropolis; lonesome places of dreadful possibilities, which could not have been more conveniently placed for the purpose of these night-hawks had they been expressly designed for them.

Travellers, who looked upon being robbed once upon a journey as the inevitable thing, very soon discovered this overcrowded state of affairs,

and resented it. Once upon a time, after the
gentry who plied their occupation on Hounslow
Heath and Finchley or Putney Commons had
taken toll of purse and pocket, travellers had gone
their way chuckling at the store of notes and gold
still safe in their boots and the lining of their
coats; but when every reckless blade and every
discharged footman or disbanded soldier took
to the road, the polite highwayman of the
recognised robbing-places had no sooner been
left behind with a " good-night to you "—mutual
good wishes and a hearty *au revoir !* from Du Vall
or one of his brethren—than the territory of an
unsuspected set of ruffians was entered; rough-
and-ready customers, who were not content until
they had got the passengers' boots off, or had
ripped up the linings of coats and waistcoats,
and then, having taken the last stiver, bade those
unhappy passengers, with a curse, begone. There
was an even deeper depth of misery—when, thus
shorn and stripped, they encountered a yet more
rascally, more provincial and hungrier crew, who
in their exasperation at getting nothing, would
sometimes resort to personal violence, to vent
their disappointment and ill-humour.

At this overcrowded period, when the ordinary
course of business failed, the highwaymen were
even known to practise upon one another, like
the Stock Exchange brokers of to-day, who,
when the public hold aloof, sharpen their wits
and fill their pockets by professional dealings

In 1758 the monotony of highway robbery

was broken by a burglary at the "Bull and Mouth" coach-office, at 3 o'clock one morning, when 47 parcels, chiefly containing plate and watches, were stolen. The booty was valued at £500. The thieves carried the parcels away in a cart, and left behind them a lighted candle, which would have burned the place down had it not been discovered in time by a coachman

This was followed in May 1766 by an incident standing out in highly humorous relief. The *Public Advertiser* in that month announced:— "A few nights ago, among the passengers that were going in the stage from Bath to London, were two supposed females that had taken outside places. As they were climbing to their seats it was observed that one of them had men's shoes and stockings on, and upon further search, Breeches were discovered also. this consequently alarming the company, the person thus disguised was taken into custody and locked up for the night. The next day he was brought before a magistrate, and upon a strict examination into matters, it appeared that he was a respectable tradesman who, having cash and bills to a large amount on him, thus disguised himself to escape the too urgent notice of the 'Travelling Collectors.'"

Turnpike Trusts at this time encouraged Sabbatarian feeling by charging double on Sundays; but "knowing" travellers sometimes travelled on that day, and submitted to that imposition as the cheaper of two evils. The one they thus escaped was the imminent risk of being molested by highwaymen and stripped of all their valuables; for those gay "Collectors," as they delighted to style themselves, did not attend to business on the Sabbath. We are not, from this, to suppose that the highwaymen were at church, or at home, reading improving literature. Not at all: they did not expect wayfarers, and so took the day off. The Sunday Trading Act for many years forbidding Lord's Day employment, prevented coaches running then, and so helped to give the hard-worked nocturnal gentlemen of the road their needed weekly rest, and ensured them from missing very much. Yet anxious travellers did sometimes go on Sundays, and risk an information. When at last the mail-coaches were started, to go seven times a week, and the Post Office itself set the example of Sunday travel, away went the high-wayman's week-ends and the travellers' respite from wayside "Stand and deliver!" The stages then plied on Sundays also

As for the mails, they were immune from attack. The Post Office early issued a warning against sending gold by them; but it did so, not from fear of the highwaymen, but "from the prejudice it does the coin by the friction." High-waymen were, in fact, little feared either by the Department or by the mail-passengers, for not only did the guard's embattled condition secure them from attack, but the Post Office introduced enactments dealing very severely with highway robbery applied to the mail-coaches. The standing

reward offered the liege-subjects of the king for
arresting an ordinary highwayman was raised
to £200 in the case of an attack on the mail,
further augmented by another £100 if within five
miles of London. Mail-coaches, by consequence,
were left severely alone by the Turpins, Aber-
shaws, and others of their kind; and it has been
said that a mail-coach, unlike the old postboys
carrying the mail-bags, was never attacked.

Although this is very likely true, it must not
be supposed that the mails were never robbed.
The distinction drawn is clear. Violence was not
shown, but robberies were frequent, often on a
sensational scale. One February night in 1810,
some unknown persons wrenched off the lock of
the hind-boot on one of the mails and made away
with no fewer than sixteen North-Country bags.
Where was the guard? Probably kissing the
pretty barmaid. Again, on November 9th, in that
same year, nine bags were stolen from a mail at
Bedford; and so frequent grew robberies of all
sorts that in January 1813 the Superintendent
of Mails was constrained to issue a warning notice
to his officials:—"The guards are desired by
Mr. Hasker to be particularly attentive to their
mail-box. Depredations are committed every
night on some stage-coaches by stealing parcels
I shall relate a few, which I trust will make
you circumspect The Bristol mail-coach has
been robbed within a week of the bankers' parcel,
value £1000 or upwards The Bristol mail-coach
was robbed of money the 3rd instant to a large

amount   The 'Expedition' coach has been twice
robbed in the last week—the last time of all the
parcels out of the seats.   The 'Telegraph' was
robbed last Monday night between the Saracen's
Head, Aldgate, and Whitechapel Church, of all
the parcels out of the dicky.   It was broken open
while the guard was on it, standing up blowing
his horn.   The York mail was robbed of parcels
out of the seats to a large amount."

Many of these robberies cited by Hasker were,
it will be noticed, from stage-coaches.   Despite
this warning note, small thefts continued.   Then,
in 1822, came the classic instance—the robbery
from the Ipswich Mail, when notes worth
£31,198 mysteriously disappeared.   A month later
the bulk of them, to the value of £28,000, was
returned, only a few, worth £3000, having been
successfully negotiated.   On the night of June 6th,
1826, seven bags were taken from the Dover Mail
between Chatham and Rainham; and in the following year a new sensation was provided by the
Warwick Mail being robbed of £20,000.

But the closing great robbery of the coaching
age was that of £5000 in notes from the "Potter"
(Manchester and Stafford) coach, October 1839.
The notes, in a parcel addressed to a bank at
Hanley, were extracted from the hind-boot when
the coach was near Congleton.

Adventures, says the proverb, are to the
adventurous; but in coaching times they befell
those who desired a quiet life, equally with
the seekers after sensation and experience.

Fortunately for the peace of mind of our grand-fathers, the startling adventure that befell the up Exeter Mail at Winterslow Hut, on the night of October 20th, 1816, was unique. The coach had left Salisbury in the usual way, and had proceeded several miles, when what was thought to be a large calf was seen trotting beside the horses in the darkness When the lonely inn of Winterslow Hut was reached, the team had become extremely nervous, and could scarcely be kept under control. At the moment when the coachman pulled up, one of the horses was seized by the supposed calf, and the others of the terrified team began to kick and plunge violently. The guard very promptly drew his blunderbuss, and was about to shoot this mysterious assailant, when several men, accompanied by a large mastiff, came on the scene; and it appeared that this ferocious "calf" was really a lioness, escaped from a travelling menagerie, and these men come in pursuit. The dog was holloaed on to the attack, and the lioness thereupon left the horse, and, seizing him, tore the wretched animal to pieces.

At length she was secured by a rope, and taken off in captivity. The leading horse was fearfully mangled, but survived, and was exhibited for a time, with great financial success, by the show-man whose lioness had wrought the mischief. When the interest had subsided, " Pomegranate " —for that was the name of the horse—was sold. He had been foaled in 1809, and was a thorough-bred, with rather too much spirit for his owner,

who had sold him out of his stable for his bad
temper. The severe work in coaches of that
period soon took the unruly nature out of such
animals, and no complaint was made of him in
his long after-career on the Brighton and Petworth
stage-coach.

This exciting episode was, of course, the
wonder of that age, and two coaching artists
made capital out of it, in the shape of very
effective plates. James Pollard was the author
of one; the other was by one Sauerweid, whose
name is not familiar in work of this kind.

Dark nights in wild country were fruitful in
strange experiences, aided, doubtless, by the
potency of the parting glass as well as by the
blackness of the night and the ruggedness of
the way. The adventures of Jack Creery and
Joe Lord, coachman and guard of the pair-horse
Lancaster and Kirkby Stephen Mail, one snowy
night, form a case in point. They had the coach
to themselves, for it was not good travelling
weather. Creery, we are told, "felt sleepy"—a
pretty way of saying he was intoxicated—and
so the guard took the reins. In driving, this
worthy, whose condition seems to have been
only a shade better than that of his companion,
wandered in the snow into a by-lane between
Kirkby Stephen and Kirkby Lonsdale, and so lost
his way. After floundering about for some time,
he aroused Creery, and their united efforts, after
alighting many times to read the signposts,
brought them in the middle of the night to a

THE LIONESS ATTACKING THE EXETER MAIL, OCTOBER 20TH, 1816

After A. Sauerweid.

village, where they were found by the aroused villagers loudly knocking at the church door, under the impression that it was a public-house. That snowstorm must have been a particularly blinding one, or the brandy at their last house of call unusually strong.

Not often was coaching history marked by such a gruesome incident as that which befell a coach on the Norwich Road. At Ingatestone a lady, who was the only inside passenger, was discovered to have died. Her son, travelling outside, was informed, but after some hesitation it was decided that the coach should proceed to its destination at Colchester. At Chelmsford, however, two ladies presented themselves as would-be passengers. Inside seats only were available, all the outsides being occupied. They were informed of the circumstances, and that they could therefore not be booked; but were so anxious to go by the coach that they overcame their natural scruples, and rode with the dead woman to the journey's end.

Of winter travelling we have already heard something, and shall hear more. How it struck one contemporary with those times we may learn from a reminiscent old traveller, who, having had much experience of old coaching methods, preferred the railway age—at least in winter. Thus he recalls some of his experiences :—

"For a day and night journey the agony was, on two occasions, so intense that, although then in my youth, and hardy enough, I was obliged to

get off the coach and sleep a night on the road ;
by which I don't mean under the hedge, but in
one of those fine old (and highly expensive) inns
that then were to be found at more or less regular
intervals along the great highways. Posting,
generally with four horses—a highly extravagant
way of travelling, but one in great favour with
those who could afford it—maintained correspond-
ingly high charges at all these houses of entertainment. It was all very well to rhapsodise
over the climbing roses, the fragrant honeysuckle
and the odorous jessamine that wreathed the
portals of the wayside inn in summer, or to become
eloquent over the roaring fire, at whose ruddy
blaze you toasted your feet in winter, but you
had to pay—and to pay pretty heavily—for these
luxuries. I will suppose that the traveller stopped
for dinner, which, if left to the landlady, generally
consisted of eels, or other fresh-water fish, dressed
in a variety of ways, roast fowl, lamb or mutton
cutlets, bread, cheese, and celery, for which a
charge of six or seven shillings was made. If
the meal took place after dark, there was an
additional item of two shillings or half a crown
for wax lights. Then, 'for the good of the house'
and your own certain discomfort, there was a
bottle of fine crusted port (probably two days in
bottle) seven shillings, or a bottle of fiery sherry,
just drawn from the wood, six shillings. To all
these charges must be added the waiter's fee of
one shilling or eighteenpence a head. 'Sleeping on
the road' absolved you from some of these costs,

but it was expensive in its own way It involved
tea or supper, chambermaid and boots, as well as
bed and breakfast. Breakfast, with ham and eggs,
three shillings, tea, with a few slices of thin
bread-and-butter, eighteenpence or two shillings;
a soda and brandy, eighteenpence.

"Once, in the depth of winter, I left Bramham
Park, the seat of George Lane-Fox, on the Great
North Road, to proceed to London, with a journey
before me of 190 miles. I was well wrapped up,
with enough straw round my feet to conceal a
covey of partridges; still, after going about 37
miles, I felt myself so benumbed that I began
to think whether it would be wise to go on, or
get off and sacrifice my fare to London. Upon
reaching Bawtry I felt more comfortable, the
guard at Doncaster having lent me a tarpaulin
lined with sheepskin; so I resolutely determined
to brave the pitiless storm of snow, now whitening
the ground.

"'Half an hour for supper,' exclaimed the
waiter, as we pulled up at the 'Crown.' Down
I got, entered the room, where there was a bright
fire blazing, devoured some cold beef, drank a
glass of hot brandy-and-water, and bravely went
forth to face the elements. By this time the snow
had increased, the wind had got up, and my heart
failed Back I rushed to the bar, ordered a bed,
and remained there for the night, finishing my
journey the following day.

"Again, in coming from Bath by a night-
coach, I was so saturated with wet and shivering

with cold that I got out at Reading, rushed to the 'Bear,' and slept there the night."

Such was the best travelling that money could buy in the days before England was—according to the coachmen — made a gridiron by the railways.

CHAPTER VII

SNOW AND FLOODS

SEVERE weather, in the shape of frosts, thunder-storms, or hurricanes, was powerless to stop the coach-service, but exceptionally heavy snowfalls occasionally did succeed in doing so for very brief intervals; and floods, although they never were or could be so general as to wholly suspend coaching, often brought individual coaches to grief.

In the severe winter of 1798-9, when snow fell heavily and continuously at the end of January and during the first week of February, several mails, missing on February 1st, were still to seek on April 27th, and St. Martin's-le-Grand mourned them as wholly lost. By May Day, however, they did succeed in running again!

Very few details survive of that exceptional season, or of that other, in 1806, when Nevill, a guard on the Bristol Mail, was frozen to death; but the records of the great snowstorm that began on the Christmas night of 1836 are very full.

Christmas Day, 1836, fell on a Sunday, and it is worth notice, as a singular coincidence in this country of only occasional heavy snowfalls, that the Christmas night of 1886, also a Sunday night, exactly half a century later, was marked by that well-remembered snowstorm which disorganised the railway service quite as effectually as that of 1836 did the coaches, and broke down and destroyed nearly every telegraph-post and wire in the land.

The famous snowstorm of 1836 affected all parts of the country, and only on two mail routes were communications kept open. Fourteen mail-coaches were abandoned on the various roads, and for periods ranging from two to ten days the travels of others ceased. The snowstorm itself continued for nearly a week The two routes remaining unconquered during this extraordinary time were those to Portsmouth and Poole, but precisely why or how they were thus distinguished is not made clear There is no doubt that the coachmen and guards on the Portsmouth and Poole Mails were strenuous men, but that quality was common to many of those engaged upon the mails. Nor can we find any favouring circumstance of physical geography to account for this unusual good fortune. On the contrary, those roads are in places exceptionally bleak and exposed to high winds, and the strong wind that on this occasion bared the heights and buried the hollows twenty and thirty feet deep in snow-wreaths was an especial feature of the visitation. Fortunately for all upon the roads—for those who laboured along them, and for those who were brought to a standstill in the drifts—the cold was not remarkably severe.

But never before, within living recollection,

had the London mails been stopped for a whole night within a few miles from London, and never before had the intercourse between the South Coast and the Metropolis been interrupted for two whole days. On Chatham Lines the snow lay from thirty to forty feet deep, and everywhere, except on the hilltops, it was higher than the roofs of the coaches. Nay, according to a contemporary newspaper account, "The snow has drifted to such an extent between Leicester and Northampton as to occasion considerable difficulty and danger. In some parts of the road passages have been cut where the snow had drifted to the depth of thirty, forty, and in some places fifty feet."

The great difficulty with which the coaches had on this occasion to contend was not merely the getting along the roads, but, as with these extraordinary depths of snow the natural features of the country were mostly obscured, of keeping on or anywhere near the road. Hedgerows were blotted out of existence: many trees had fallen under their snowy burdens, and it was not unusual, when at last the snowed-up mails were recovered, to find them strayed far from their course, and in the middle of pastures and ploughlands.

Snowstorms produced curious travelling experiences. It was this great occasion that effectually blocked all the up night coaches for two days at Dunchurch, on the Holyhead Road, and so succeeded in bringing together a party not unlike those weatherbound travellers who in Dickens' Christmas stories gather round the hearth, and, comforting themselves with many jorums of punch, tell dramatic stories. One party crowded the " Dun Cow," another the " Green Man." Among the coaches were the Manchester " Beehive " and the " Red Rover." The first morning of their enforced leisure they—coachmen, guards and passengers—made up a poaching party, with two guns among sixteen of them. Jack Goodwin, guard of the " Beehive," was the only fortunate sportsman, and shot a hare. In the evening a dancing party was held at the " Dun Cow " at the suggestion of the landlord, who invited some friends, and the next morning Goodwin turned wandering minstrel, taking with him a chosen few to help in chorus. Wandering along the Rugby Road, they were entertained at the farmhouses with elderberry wine and pork pies. Another pleasant evening, and they were off the next morning for London

Floods were infinitely more dangerous than snowstorms, and the Great North Road, between Newark-on-Trent and Scarthing Moor, was particularly subject to them, the Trent often, and on the very slightest provocation of rain, flooding many miles of surrounding country It was here, and on these occasions, that the outsides had the better bargain of the two classes of travelling, for they kept their seats without fear of being drowned, while the insides went in constant terror of a watery death, and often only escaped it by the

pitiful expedient of standing on their seats and so
—keeping the doubled-up attitude this necessity
and the lowness of the roof imperatively demanded
—remaining until the levels were passed and the
dry uplands reached again.

In August 1829, when extraordinary floods
devastated a great part of Scotland, a stirring
episode occurred in connection with them and the
mail-coach running through Banff. The tradition
that his Majesty's mails were to be stopped for
nobody and hindered by nothing on the road was
a very fine and fearless one, but it was occasionally
pushed to absurd lengths, and hideous dangers
provoked without reasonable cause. This episode
of the Banff and Inverness Mail is a case in point.
The mail of the preceding day had found it im-
practicable to go by its usual route, and so took
another course, by the Bridge of Alva. It was
therefore supposed that the mail following would
adopt the same plan; but what was the astonish-
ment of the good folk of Banff when they perceived
the coach arrive, within a few minutes of its usual
time, at the farther end of the bridge that crosses
the River Dovern. The people, watching the
eddying floods from the safe vantage-point of their
windows, strongly dissuaded the guard and coachman from attempting to pass, the danger being
so great; but, scouting the idea of perils to be
encountered in the very streets of the town, those
foolhardy persons drove straight along the bridge
and into a street that had been converted by the
bursting of the river-bank into the semblance of

a mountain torrent. When the furious current caught the coach, it was instantly dashed against the corner of Gillan's Inn, and the four animals swept off their legs. They rose again, plunging and struggling for their lives, and a boat was pushed off, with men eager to free the poor animals from their harness, to enable them to swim away; but it was not possible to save more than one. The other three were drowned.

By this time the coach, with coachman and guard, had been flung upon the pavement, where the depth of water was less; and there the guard was seen, clinging to the top, and the coachman hanging by his hands from a lamp-post, regretting too late the official ardour and slavery to tradition that had wrought such havoc. When, for humanity's sake, as well as to secure the mail-bags, a boat came and rescued them, they were not suffered to depart without much Aberdonian plain-speaking on the folly that had nearly cost them their lives and endangered the correspondence of the good folks of the ancient burgh of Banff.

There were no passengers on this occasion, but we are not to suppose that, had there been any, they would have received much consideration. The mail would probably have been driven on, just the same. The official attitude of mind towards them may be judged from the wintry horrors encountered by the Edinburgh to Glasgow Mail in March 1827. It became embedded in the snow near Kirkliston, and the guard, riding one

MAIL-COACH IN A SNOW-DRIFT.

After J. Pollard.

horse and leading another loaded with the bags,
set off for Glasgow; while the coachman, with the
other horses, set off in the opposite direction to
secure a fresh team, pursued by the entreaties of
the four terrified passengers, beseeching him to
use all diligence and return soon. There, on a
lonely road, immovably stuck in huge snowdrifts,
they remained throughout a bitter night, made
additionally miserable by one of the windows being
broken. It was not until nine o'clock the next
morning that the coachman returned, with another
man, but only two horses. Having loaded them
with some luggage and parcels, he was, with a
joke upon his lips, leaving the passengers to shift
for themselves, but was compelled to take one
who had fallen ill. The remaining three extricated
themselves as best they could.

On September 11th, 1829, a month later than
the watery adventures at Banff, the Birmingham
and Liverpool Mail had an unfortunate experience
at Smallwood Bridge, near Church Lawton, a point
where the road is crossed by an affluent of the
River Weaver. Unknown to those on the mail,
the flooded stream had burst the arch of the
bridge, and when the coach came to the spot,
along a road almost axle-deep in water, it fell into
the hole and was violently overturned. Of the
three inside passengers, only one escaped. He
was an agile young man, who broke the window
and so extricated himself. The horses were
drowned, but the coachman was fortunate enough
to be washed against a tree-stump as the river

hurried him along at six miles an hour. The force of this happy meeting nearly stunned him, but he held on, and eventually found his way ashore. The guard was saved in a similar manner. Accidents almost forming parallels with this were of frequent occurrence, and a seasoned traveller exclaimed. "Give me a collision, a broken axle and an overturn, a runaway team, a drunken coachman, snowstorms, howling tempests; but Heaven preserve us from floods!"

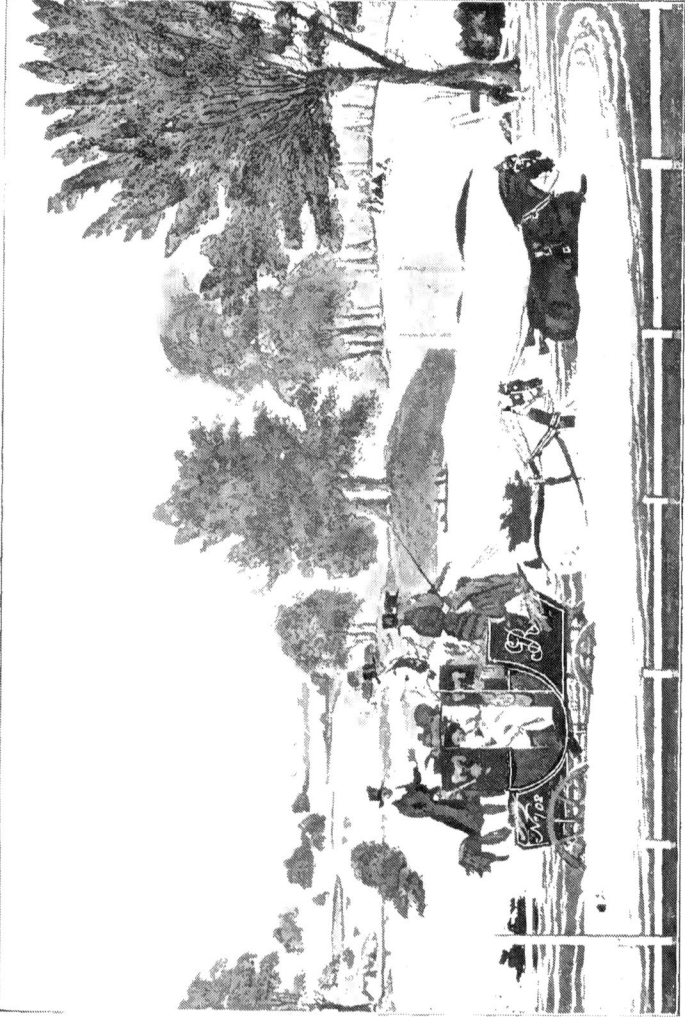

MAIL-COACH IN A FLOOD.

Aft. J. Pollard.

CHAPTER VIII

THE GOLDEN AGE, 1824—1848

IT was "golden" chiefly from the sportsman's point of view, and in the opinions of those who found a keen delight in the perfection of coach-building and harness-making, in the smartness of the beautiful horses, and in the speed attained. From the sordid view-point of the profit-and-loss account, although this was the age in which Chaplin and a few others made their great fortunes, it was a time when the high speed and other refinements of travelling made the path of the coach-proprietor a thorny and uneasy one, often barren of aught but honour. "You are 'in it,' I see," said a proprietor who himself had been severely bitten in this way, and had left the business, to a coachman who, like many of his fellows, had long cherished an ambition to become a coach-master, and had just acquired a share: "you are 'in it.' Take care how you get out of it." One of the prominent men in it—Cooper, who ran a good line of coaches on the Bath Road—found himself at last in the Bankruptcy Court, and many smaller men appeared in the same place. The greatest increase of cost was in the item of horses. In earlier times the stock had lasted for years, despite the long stages and harder pulling; but in this period of good roads and short stages, when, all things being equal, a team should have lasted longer, the great coach-proprietors found it necessary to renew their stock every three years. Chaplin's method of doing this was to replace one-third of his horses every year.

It is not to be supposed that the horses thus disposed of were always broken down or worn out by their three years of strenuous exertion in the fast coaches They had only lost those agile qualities necessary for that use, and, finding purchasers among farmers and country tradesmen who had no occasion to gallop at eleven miles an hour, lived very comfortably, grew sleek and fat, and must often, from roadside paddocks, have beheld their successors slaving away in the fast coaches; finding much satisfaction in their own altered circumstances Coachmen at this time usually drove between thirty and forty miles out, and then took the up coach back, perhaps more than half a day later With such an arrangement the horses had the same driver, and it was generally found that they worked much better in such cases. The coachman's responsibility for their condition was also undivided, and the proprietor was easily able to weed out from his coachmen those who lingered at the changes and made up the lost time on the road, to the distress of their teams. It was Chaplin who made it known, by all the rigorous language at his

command, that any one of his coachmen found
in the possession of one of those instruments of
torture, resembling a cat-o'-nine-tails, for punishing horses, and known as a "short Tommy"
would be instantly dismissed   Chaplin's direct
influence and interests may be said to have
described a radius of from forty to fifty miles
from London, and within that circle the "short
Tommy" was therefore but seldom seen.   One
historic occasion there was, however, when such
an object did most dramatically present itself
before Chaplin, who chanced to be at a wayside
inn when one of his coaches pulled up to change.
On the roof was a warder with two convicts.
As the coachman, with much deliberation, lowered
himself from his box to the ground, the "short
Tommy" he had been sitting on fell in front
of the windows, and as it lay there attracted
the eagle eye of that great coach-proprietor, who,
sternly bent upon executing justice upon the
offender, strode forth.   The coachman, dismayed,
saw his employer and the forbidden instrument
at once, in one comprehensive, understanding
gaze, but he was a resourceful man, and handed
it to the warder, telling him, with a portentous
wink and a warning jerk of the head, that he
had dropped something.   That worthy, entering
into the spirit of the deception, accepted the
pretended cat-o'-nine-tails, and the coachman
breathed freely again.

The days of ten- or eleven mile- stages, just
at this time faded away, gave a horse one stretch

of so many miles a day; but in the fast coaches
of the newer age they ran, as we have seen, out
and home, six or seven miles each way. It was
to the very last a disputed point whether it
was better for a horse to do his ten or eleven
miles and have done with it for the day, or to
do his two shifts of six or seven Many coachmen who could not depend upon their horse-
keepers objected to two sweats a day; but this
division of work was a decided advantage to the
horses, if well tended, and in such cases they
had the advantage of sleeping at home every
night. The number of horses kept for one of
the fast coaches of this Augustan age would
have astonished the pioneers of coaching; one
horse for every mile travelled was the establishment kept up. Slow coaches could do with fewer.

The average price paid for a coach-horse at
this period was £30, but some were acquired for
a mere trifle, owing to their being vicious or
unmanageable in private hands. The private
owner's dilemma was the coach-proprietor's opportunity. It mattered little to him what defects
of temper a horse possessed so long as he was
sound in wind and limb. For the rest, a little
discipline, harnessed with three others, all subject
to the rule of those very able disciplinarians, the
coachmen, quickly sufficed to bring such an
animal to reason There were thus some very
queer animals drawing the coaches in these last
years.

Some were purchased with a doubtful title.

In such a case, to prevent his being recognised and claimed, the horse would be worked on the night mail.

The coachman's ideal was a team matching in colour, but few proprietors ever aimed at such perfection. The cost was great, and nothing, save the gratification of the eye, was gained.

With these business details the travelling public had no concern, and it was only the box-seat passengers who learnt the history of some of these cheap acquisitions from private stables. The box-seat passenger was generally a sporting character, aspiring to that companionship with the coachman from his love of horses and driving, but it naturally often happened that some stolid person, whose only desire was to be carried safely and who took no interest in driving, found himself perched on that place of honour. When such an one became the unwilling confidant of the coachman he was apt to hear some nerve-shaking things. "See that 'ere near wheeler?" said one John. "Run away vith a old gennelman last veek, he did; broke his neck; friends just goin' to shoot 'im; guv'nor gave couple o' suvrings for 'im, and 'ere 'e is. 'Ope we shan't be upset!" The nervous passenger effected an exchange for an inside place with a sporting passenger at the next stage—which was precisely the result anticipated by the coachman.

At this time, when the fast day-coaches were in every respect as well appointed as a gentleman's private drag, it was the keenest ambition of every dashing young traveller to occupy this box-seat—an ambition generally satisfied by putting in an early appearance at the starting-point and tipping the head yard-porter, who thereupon placed a rug or some stable cloths on it. These tips were not, as generally supposed, the coachman's perquisite. His turn came later on, down the road.

The yard-porter was a much more important official than the present generation might suppose, and in busy yards, such as those of the " Bull and Mouth " or the " Swan with Two Necks," his weekly income from tips probably amounted to £5. or more. Nor was he merely the man with a pail of water, a broom and a pitchfork conjured up mentally by the sound of his title; his was an important department, and himself the ruler of many subordinates, whose duties ranged from grooming and bedding-down the horses and cleaning the stables to washing the coaches and cleaning and polishing the harness and metal-work.

At this period the public found themselves swiftly flying where they had formerly slowly and laboriously crawled, and generally compared ancient travelling with modern, greatly to the advantage of modern times. But if the coach-proprietors who were at such pains to compete with one another in establishing these swift and well-appointed coaches were of opinion that in so doing they were earning the admiration of the entire travelling public, they were very soon

undeceived, and those weaker brethren who could
not command the influence and the capital by
which only could a fast coach be appointed
and established, found that, after all, there was
no immediate prospect of their being run off the
road, and that a considerable section of the public
actually preferred to travel in slow coaches, and
would by no means consent to be whirled through
the country at eleven miles an hour, with only
hurried intervals for meals "The art of travelling,"
said an anonymous writer in 1827, " has undergone
great alterations in the course of the last thirty
years; these are not altogether improvements."
One of these changes for the worse, in the opinion
of this unknown scribe, was that in the thunder
of ten miles an hour there was no opportunity
for conversation.   That must be a powerful tongue,
he thought, which could make itself heard amid
the reverberations of such incessant and intem-
perate whirlings.   He could not help looking back
with some regret to the good old times when
five or six miles an hour was the utmost speed.
Then there was something sober and sedate in
the fit-out and the set-out.   All the faces in the
inner-yard were so grave and full of importance,
and there was some seriousness in taking leave.
(Good reason, too, for such gravity and seriousness,
think we of later ages.)   How scrupulous and
polite were the inside passengers, in making
mutual accommodation of legs and arms, band-
boxes, sandwich-baskets and umbrellas!   Then,
too, says this delightful snob, there was some

difference between the inside and the outside passengers: the gentlefolks within were not confounded with the people on the outside. Distinctions were then better observed, and preserved. Older stage-coach conversation, he continued, was apt to be conducted with caution, for a false opening might make an ill companion on a long journey. So approaches were made skilfully, and with deliberation. A man was thought excessively forward and talkative if he had got into politics before he had well cleared the outskirts of London, and the first half-hour was generally occupied with the light skirmishings of talk, with reconnoitrings of one's opposite neighbour's countenance, and a variety of all-round questions and answers put and returned merely to ascertain how far the passengers were to be companions. These had to be framed with the utmost discretion. With what vivacity and air of pleasant expectation would one then ask an agreeable-looking person, "Are you going all the way to Toppington?" or, on the other hand, if the inside had its full complement of six, how carefully, and with what a discreetly modulated voice, in order to avoid all suspicion of wishing a speedy riddance, one would ask the same question of an unduly stout person, who occupied much more than his or her share of room.

The best conversational opening was considered to be, "Well, we are now off the stones. What a beautiful morning! How charming the outskirts of town! Pray, does not that house belong to —— ?"

Going up-hill one walked, to ease the horses, insides and outsides then equal; the insides, greatly condescending, holding converse with the occupants of the roof, always, however, with the strict understanding—no less strict if not mentioned —that this gracious act must not be taken advantage of by those outsiders claiming acquaintance when the coach stopped at the inns, where this all-important difference in caste was recognised by distinct eating apartments being provided.

Those were the good old days, according to this critic, when these customs were strictly observed, and when there was not only time to eat, but almost to digest at coach-dinners and breakfasts; when, too, there were generally a few minutes to spare while the horses were being got ready, so that the passengers could wander round the town and copy any curious epitaphs for the *Gentleman's Magazine*, or do a little shopping.

Coachmen were of somewhat similar opinions. "Lord! sir," said Hine, coach-proprietor and coachman on the Brighton Road, in 1831, who was, much against his will, obliged to accelerate his coaches in order to keep pace with newcomers, but did not relish the necessity, " we don't travel half so comfortably now as we used to do. It is all hurry and bustle nowadays, sir—no time even for a pipe and glass of grog." Not comfortable for the coachmen, who sadly missed their wayside, and often wholly unauthorised, halts.

Cobbett, surly though his nature was, could

not withhold admiration when noticing these
latter-day coaches. "Next to a fox-hunt," he
says, "the finest sight in England is a stage-coach
just ready to start A great sheep- or cattle-fair
is a beautiful sight; but in a stage-coach you
see more of what man is capable of performing.
The vehicle itself; the harness, all so complete
and so neatly arranged, so strong, and clean,
and good; the beautiful horses, impatient to be
off; the inside full, and the outside covered, in
every part, with men, women, and children,
boxes, bags, bundles; the coachman, taking his
reins in one hand and his whip in the other,
gives a signal with his foot, and away they go,
at the rate of seven miles an hour—the population
and the property of a hamlet. One of these
coaches coming in, after a long journey, is a
sight not less interesting. The horses are now
all sweat and foam, the reek from their bodies
ascending like a cloud. The whole equipage is
covered, perhaps, with dust and dirt. But still,
on it comes, as steady as the hand of the clock.
As a proof of the perfection to which this mode
of travelling has been brought, there is one coach
which goes between Exeter and London, whose
proprietors agree to forfeit eightpence for every
minute the coach is behind its time at any of
its stages; and this coach, I believe, travels eight
miles an hour, and that, too, upon a very hilly,
and at some seasons a very deep, road."

Yes, but had Cobbett written in still later
years, he would have found the "Quicksilver"

LATE FOR THE MAIL.

After C. Cooper Henderson, 1848.

attaining, between the stages, a speed of nearly 12 miles an hour, and an average speed, including stops, of 11 miles, while a quite ordinary performance with the Shrewsbury "Wonder" was 158 miles in 14 hours 45 minutes, including stops on the way totalling 80 minutes. This gives a net average speed of a little over 11½ miles an hour. The Manchester "Telegraph" and other flyers made equally good performances. The "Tantivy," one of the most famous of coaches, did not equal these feats.

The "Tantivy," London and Birmingham coach, was started in 1832. It left the "Blossoms" inn, Lawrence Lane, at 7 a.m., and was in Birmingham by 7 p.m. The distance, by the route followed, through Maidenhead, Henley, Oxford, Woodstock, Shipston-on-Stour, and Stratford-on-Avon, was 125 miles, and, deducting one hour for changing and refreshing, the speed was only slightly over 11 miles an hour. This coach derived its name from the old word "Tantivy '— an imitative sound as old as the seventeenth century, and often used in the literature of that time, supposed to reproduce the note of the huntsman's horn, and conjuring up ideas of speed. For Cracknell, the most famous of the coachmen of the "Tantivy," who once drove the 125 miles at one sitting, and generally drove it between London and Oxford, the "Tantivy Trot," quoted elsewhere in these pages, was written. Harry Salisbury drove between Oxford and Birmingham   Among its other coachmen was

Jerry Howse Costar and Waddell, of Oxford, horsed the "Tantivy" between Woodstock and London, and Gardner, of Stratford-on-Avon, part-horsed it onwards, not wholly to the satisfaction of Salisbury, who used to declare that the team out of his yard was worth about £25 the lot, and that they had once belonged to Shakespeare.

Competition in speed led naturally to rivalry in the building, upholstering, and general appointments of the coaches. Sherman's Manchester "Estafette" was a splendid turn-out, holding its own against many rivals in the last years of the coaching age. Inside was a time-table elegantly engraved on ivory, showing all towns, distances and intermediate times, illuminated at night by a reflector lamp It was at this time seriously proposed to light the coaches with gas, with the double object of securing better lighting and effecting a saving on the very heavy bills for oil consumed on the night coaches. The idea was generally abandoned when it was found that the gas tanks would be very heavy and that they would take up all the room in one of the boots, generally reserved for luggage. Coachmen and guards, too, professed anxiety lest they, sitting directly over the fore and hind boots, should be blown up But, before the project was finally abandoned, it was fully proved that it was practicable, and in January 1827 the Glasgow and Paisley coaches were lit with gas, much to the amazement of the country folk. "Guid Lord, Sandy," said an old woman to her

husband, "they've laid gas-pipes all the way
frae Glasgae Cross to Paisley!" But they had
done nothing of the sort; the gas was carried, as
already indicated, in a reservoir stowed away in
the front boot

Competition having already raged around the
question of speed, and having introduced unwonted luxuries in travelling, turned next to
the more deadly form of rate-cutting. In 1834
the coach-proprietors on three great routes were
engaged in this game of Beggar-my-neighbour.
In that year the fares to Birmingham, Liverpool,
and Manchester fell to less than half their former
price, and it was possible to travel to Birmingham
for 20s inside and 10s. out, or to Liverpool or
Manchester for 40s. inside or 20s. out. They had
little chance of being raised again, for, by the
time the weaker men had been crushed out of
existence, the railways were threatening the
whole industry of coaching.

But reducing the fares by one-half was not
always the last word in these bitter contests.
There was a period on the Brighton Road when
one might have been carried those 52 miles in
6 hours for 5s., with a free lunch and wine at
the end of the journey and your money returned
if the coach did not keep its time. The "Golden
Age," indeed!

At this period, when the long-distance coaching
business was so severely cut up, those proprietors
who served the districts surrounding London did exceedingly well   Coaching annals are almost silent

on the subject of these suburban coaches, for, being drawn by only two horses, they were regarded by the four-in-hand artists with contempt. It has thus, in the absence of available information, often puzzled inquiring minds in the present generation to understand how the heavy passenger traffic was conducted between London and the out-lying towns and villages within a radius of twenty miles. Those districts were served by these "short stages," as they were called—coaches drawn by two horses, and making two or more journeys each way daily. There was an incredibly large number of these useful vehicles, which were in relation to the mails and fast long-distance coaches what the suburban trains are to the expresses in our own day. The ordinary coach-proprietors had rarely anything to do with these conveyances, which came to and set out from a number of obscure inns and coach-offices in all parts of the City and the West End.

One of these short stages is mentioned in *David Copperfield*, where David's page-boy, stealing Dora's watch and selling it, purchases a second-hand flute and expends the balance of his ill-gotten gains in incessantly travelling up and down the road between London and Uxbridge. Evidently a lover of the road, this larcenous page-boy. Most boys in buttons (and certainly the typical page-boy of the typical farce) would have expended the plunder in pastry or tobacco. This particular specimen, the diligent Dickens-reader will remember, was taken to Bow Street on the completion of his fifteenth journey, when four shillings and sixpence and the second-hand flute— which he couldn't play—were found upon him. If we were contemplating an examination-paper on *David Copperfield*, with special reference to prices and social life early in the nineteenth century, we might put the following poser :— " State the average price obtainable on the average lady's gold watch, and, deducting the purchase price of a second-hand flute, deduce from the resulting sum, and from the facts of the boy having made the journey fifteen times and still being in possession of four-and-sixpence, the cost of a single outside journey between London and Uxbridge."

The fare was, as a matter of fact, half a crown. There were no fewer than seven short stages between London and Uxbridge daily, each making two journeys. What of those London inns whence they started ? Where are they now ? Echo does not answer " where ? " as she is commonly said to do, because it is not in the nature of echoes to repeat the first word of a sentence. No ; echo merely reverberates " now ? " with a questioning inflection.

The " Goose and Gridiron," whose proper name was originally the " Swan and Harp," in St. Paul's Churchyard, was one of these starting-points. From the same inn went the Richmond and many other suburban stages. That old inn was de- molished about 1888 The " Boar and Castle and Oxford Hotel," No. 6, Oxford Street, was another house of call for the Uxbridge stages. It vanished long ago, and those who seek it will only find on its site the Oxford Music Hall and Restaurant—bearing a different number, for the street was re-numbered in 1881. The " Boar and Castle " was a large, plain, stucco-fronted house, with its name writ large across the front in raised letters

As for the " Old Bell," another of these starting-points of the Uxbridge coaches, it was pulled down in 1897. It stood on the site of Gamage's, in Holborn, opposite Fetter Lane. Of another Uxbridge house, the " Bull," a few doors away, the sign, the figure of a ferocious black bull, very properly chained and fastened by a secure girth, still exists on the frontage, but " Black Bull Chambers," a set of grimy modern " model " dwellings, now occupy the coach-yard. The " Bell and Crown," afterwards " Ridler's," next Furnival's Inn, has been swept away to help make room for an extension of the Prudential Assurance Offices, and the " New Inn," 52, Old Bailey, has given place to warehouses and the premises of a firm of wholesale newsagents. Away westward, the Uxbridge and other short stages called at the " Green Man and Still," at the corner of Argyll Street, Oxford Circus, and at the " Gloucester Warehouse," near Park Lane The last-named was rebuilt forty years ago, but the " Green Man and Still " was only demolished in February 1901.

The time taken over the eighteen miles between the City and Uxbridge was three hours. To Richmond in 1821, when short stages ran

THE SHORT STAGE.

After J. Pollard.

frequently from five different inns, the time was an hour and a half. As many as fourteen coaches ran to that town in 1838, most of them making six journeys a day. Shillibeer and his omnibuses, introduced in 1829, had by that time rendered the exclusive short-stages old-fashioned, and they were gradually replaced by the more commodious and popular vehicles, whose occupants were in turn looked down upon by the short-stage passengers, just as *they* had been despised by the four-horse coaches.

CHAPTER IX

COACH-PROPRIETORS

NONE among the servants of the public earned their living more hardly, or took greater risks in the ordinary way of business, than the coach proprietors. It was a business in which the few—the very few—became rich, and the majority lived a strenuous life, with empty pockets at the end of it. It was very truly said of them, as a class, that they lived hard, worked hard, swore hard, and died hard. To this was sometimes added that they held hard, by which you are to understand that what money they *did* succeed in getting they grasped tightly. This last was, however, by no means a characteristic of the majority, who very often dissipated what they had made by successful ventures on one road by disastrous competition on another. There was never a more speculative business than that of a coach proprietor, and never one so cursed with insane competition. Why embittered rivalries of this kind should have been more common on the road than in any other line is only to be explained by the hypothesis that a certain element of sporting emulation entered into it, and a kind of foolish pride that impelled a man to put and keep a line of coaches on a road to "nurse" a rival, not always with the hope of earning a profit for himself, but with the idea of cutting up another man's ground.

The most outstanding figure among coach-proprietors was that of William Chaplin. He towered above all his contemporaries in the magnitude of his business, and was, when railways came to destroy it, first among those few who saw the folly of opposing steam, and were both acute enough and sufficiently fortunate to reap an additional advantage from the new order of things, instead of being ruined by it, as many less fortunate and less far-seeing men were.

William James Chaplin — to give him his full baptismal name — was born at Rochester in 1787, the son of William Chaplin, at that time a coachman and proprietor in a small way of business on the Dover Road. Shortly after that date it would appear that the elder Chaplin extended his operations, and became a coach-master on a considerable scale on some of the main roads leading out of London. However that may have been, certain it is that his son was a practical coachman, and thoroughly versed in every detail of driving and stabling, as well as buying horses. To this intimate acquaintance with the conduct of a coach and of a coaching business, as greatly as to his own native shrewdness, he owed the extraordinary success that attended him. His centre of operations was at the "Swan with Two Necks," in Lad Lane, where he succeeded William

Waterhouse, who had been established there as a mail-contractor since 1792. He it was who, perhaps in imitation of the Mail-coach Halfpenny dedicated to Palmer, issued the curious copper token pictured here. It is quite in accord with the general fragmentary character of the records of these not so remote times that nothing survives by which we may state the year when Chaplin succeeded Waterhouse at the "Swan with Two Necks," but it was probably about 1825. In addition to this yard, he acquired in the course

MAIL-COACH HALFPENNY ISSUED BY WILLIAM WATERHOUSE.

of time those of the "White Horse," Fetter Lane, and the "Spread Eagle" and "Cross Keys," Gracechurch Street, together with the "Spread Eagle" West End office, in Regent's Circus, with the proprietorship of several hotels. Unlike most coach-proprietors, who restricted their operations to one or two roads, Chaplin's coaches went in all directions, and he owned large stables at Purley on the Brighton Road, at Hounslow on the Western roads, and at Whetstone on the great road to the north. The "Swan with Two

WILLIAM CHAPLIN
From the painting by Frederick Newnham.

Necks," was, when he acquired it, a yard extremely awkward of approach, being situated in a narrow lane, and inside a low-browed entrance that taxed the ingenuity of the coachmen to pass without accident. Once inside, you were in one of those old courtyards without which no old coaching inn was complete. Three tiers of galleries ran round three sides of the enclosed square, which, from the creepers that were trailed round the old carved wooden posts or depended from the balusters, and from the flower-boxes that decorated the windows, was a very rustic-looking place. Chaplin had not long settled himself here before he constructed underground stables beneath this yard, where some two hundred horses were stalled; but the place remained, otherwise unaltered, until about 1856, when all the buildings were demolished, and he set himself to raise on their site the huge pile of buildings that now fronts partly on to Gresham Street and partly to Aldermanbury. It was one of his last works, and was, of course, undertaken long after the coaching age had become a thing of the past, being, indeed, intended for the head-quarters of the carrying business that had in the meantime come into existence. It is of somewhat curious interest to note that, although the great gloomy pile of unadorned brick bears not the slightest resemblance to the ancient coaching inn, yet a courtyard survives, and railway vans manœuvre where of old the mails arrived or set forth.

In 1838, when his coaching business had reached its full height, Chaplin owned or part-owned no fewer than 68 coaches, with 1,800 horses. Twenty-seven mails left London every night, and of these he horsed fourteen on the first stages out of and into town. The annual returns from his business were then put at half a million sterling

At this critical period he resided at an hotel he owned and managed in the Adelphi, where he worked literally day and night, supervising the general affairs of his vast business, and yet finding time for correcting details. Those coachmen who thought themselves secure from observation in the midst of all these extensive operations wofully deceived themselves. They had to reckon with one to whom every detail was familiar — who had driven coaches himself, and was thoroughly informed in the opportunities that existed in the stables and on the road for cheating an employer. He knew the measure of every corn-box, and was cognisant of the "shouldering" of fares and "swallowing" of passengers that continually went on. For the guards thus to pocket the short fares, not entering them on the way-bill, afterwards sharing them with the coachman, was a practice that went back to the very early days of coaching, and not only lasted as long as coaching itself, but survived in a somewhat altered form on omnibuses until the introduction, in recent years, of tickets and the bell-punch. It would have been impossible for coach-proprietors to end

THE CANTERBURY AND DOVER COACH, 1830.

After G. S. Veigma.

this practice without raising the wages of their servants, and thus they were obliged, so long as the coachmen and the guards performed their " shouldering " and " swallowing " discreetly, to allow it to continue   The practice was, indeed, a very lucrative one to those chartered peculators, who made a great deal more out of it than they would in the substitution of higher wages and a better code of morals.   Like the omnibus-proprietors until recently, coach-masters were content so long as their takings reached a certain average sum, and it was only when they fell below that figure, or when a fare was " shouldered " or a passenger " swallowed " before their very eyes, that trouble began.   Chaplin could thus afford to give the toast, as often he did give it, at festive gatherings of coachmen and guards, " Success to ' shouldering,' but " (with a peculiar emphasis) " do it well ! "—or, in plainer speech, " don't get found out ! "

Stories with Chaplin for a central figure were, of course, plentiful down the road.   Stable-folk told how one of their kind, who had been requisitioning the contents of the corn-bin to an extravagant extent, going to it with sack and lantern one night when all was still, lifting the lid, found Chaplin himself snugly waiting within, who promptly arose in his wrath, and, to the accompaniment of a picturesquely lurid eloquence of which he was an undoubted master, dismissed him instanter.   The fame of that exploit must have saved Chaplin much in forage.

Although in his after-career as Member of Parliament he was a silent representative, he could be eloquent in various ways. He had, as already hinted, the direct and forcible method in perfection, and yet could suit his style to all requirements. Coachmen, indeed, found him much more dangerous in his suave and polite moments, and much preferred to be sworn at and violently attacked, for his polite speeches generally had a sting in their tail, and earned him, among the brethren of the road, the descriptive, if also disrespectful, nickname of " Billy Bite-'em-Sly."

The portrait of him shows a physiognomy altogether unexpected, after hearing these tales. One perceives rather a delicate and refined face than that mentally pictured, and it is only in the piercing eyes that his energy and determination are clearly seen.

Chaplin's coaches were easily to be distinguished along the roads, not only by the device of the " Swan with Two Necks " painted on them, or later, in addition, by those of a " Spread Eagle," " Cross Keys," or a " White Horse," as those inns came under his control, but by their colours, which were red and black—black upper-quarters and fore and hind boots, and red under-parts and wheels.

His coaching business gave employment to two thousand people, and included a horse-buying and veterinary department, under the control of James Nunn, who was accustomed to procure the greater

JAMES NUNN, HORSE-BUYER AND VETERINARY SURGEON TO WILLIAM CHAPLIN.

After J. F. Herring.

number of the coach-horses from Horncastle Fair
J F. Herring has left an excellent equestrian
portrait of this indispensable personage.

Chaplin horsed the quickest mails out of
London: the Devonport, the New Holyhead, the
Bristol, and five other West-country mails starting
from Piccadilly. Passengers who had booked
from his City offices were carried to this point
by omnibuses he established, and the mails were
conveyed, with the guards, in two-wheeled mail-
carts from the General Post Office. In the great
number of coaches he ran there were, of course,
included some of the very best. His were those
famous coaches, the Manchester "Defiance," a
rival of Sherman's even more famed Manchester
"Telegraph," the Birmingham "Greyhound," the
Cambridge "Telegraph," Liverpool "Red Rover,"
Bristol "Emerald," Cheltenham "Magnet," and
many others doing their ten miles and more an
hour. He also had half-shares in the brilliant
"Tantivy," London and Birmingham, the "Stamford Regent," the Southampton "Comet," and
others.

The signs of the times, so patent to outsiders
from 1830 and onwards, but generally hid from
the vision of those most interested, were not
unheeded by this remarkably shrewd business
man, who, like his contemporary, Joseph Baxendale, had the power of seeing things and the
possible future trend of affairs from an impersonal
and unprejudiced point of view. He, above all
other coach-proprietors, was deeply interested in

the continuance of the old order of things, and it would not have been remarkable had he brought himself to the illogical conclusion that, because he was so interested, the old order must, could, should and would be maintained. Many other coach-proprietors *did* arrive at such a conclusion, not, of course, by process of reasoning, but by force of being habitually engaged in a business that pre-judiced their minds against steam and machinery. Their first instincts of scorn for anything that should presume to replace the horse effectually blinded them to the reality of the coming change

Chaplin early decided that coaches must go, and that the proper policy was to make allies of the railways in early days, while they were not so sure of their own success, and would be sub-stantially grateful for any helping hand. He and Benjamin Worthy Horne agreed with the London and Birmingham to be their very good friends in this matter, and not only withdrew all competitive coaches as the line advanced towards completion, but aided the railway in those months when a gap in the line between Denbigh Hall and Rugby cut the train journey in two. Between those two points their coaches filled the unwontedly humble position of feeders and go-betweens to the railway. The price of this amiable attitude was a share with Pickford & Co. in the goods and parcel cartage agency for the line, to the exclusion of all others This monopoly, as Chaplin had foreseen, was an initially valuable one, and certain to constantly increase, side by side with the growing trade and

mileage of the railway itself He sold most of
his coaches—who were those rash persons, greatly
daring, who bought coaches in those last days?—
and realised everything except what was considered
necessary to start the new firm of Chaplin &
Horne, carriers, and to carry on the branch
coach-services on routes not yet affected by the
rail. Having thus converted his fortune into
hard cash and deposited it for the time being
in the bank, the next consideration was what to
do with it. All the preconceived ideas of investment were being uprooted, and railways, which
offered many chances to the capitalist, were not
in those times bracketed with Government securities as safe. Even supposing railways in general
offered inducements, those were the days when
they were not merely unproved, but when few
had advanced beyond the point of obtaining their
Parliamentary powers. They were, in fact, little
but projects on paper. With these problems to
consider, Chaplin did a singular thing. Leaving
his fortune on deposit, he went away and utterly
secluded himself in Switzerland for six weeks, to
to debate within himself this turning-point in a
career. He was now fifty-one years of age, and
might well have been content with what he had
accumulated, and with the prospects of the new
firm. With the advantages he had already secured
he could have enjoyed a leisured life ; but he took
the decision to embark a large portion of his cash
in the London and Southampton Railway, then
under construction and very much under a cloud

of depreciation. He aimed at becoming a director on that line, and had that desire speedily gratified, being further appointed Deputy Chairman in 1839. By 1843 he had succeeded to the chair, and, with one interval, remained Chairman of what became the London and South-Western Railway until 1858, when ill-health compelled his resignation. He had the satisfaction of seeing his belief in the future of that railway assured. He was also a director of the Paris and Rouen, the Rouen and Havre, and the Rhenish Railways; Sheriff of London, 1845-6; a Member of Parliament for Salisbury, 1847-57; in politics an advanced Liberal. He died at his residence, 2, Hyde Park Gardens, on April 24th, 1859, in his seventy-second year, leaving property to the value of over half a million sterling, including a quarter share in the firm of Chaplin & Horne William Augustus Chaplin, the eldest among his eight sons and six daughters, succeeded him in the conduct of that business, and died, also in his seventy-second year, at Melton Mowbray, October 9th, 1896.

Benjamin Worthy Horne, whose chief place of business was the "Golden Cross." Charing Cross, succeeded his father, William Horne, in 1828. William Horne, who was born in 1783, was originally a painter, but followed that trade only a few years after his apprenticeship had expired. He had at an early age married Mary Worthy, daughter of Benjamin Worthy, a wealthy wheelwright in Old Street, and in 1804 his eldest son, Benjamin Worthy Horne, was born. This

WILLIAM AUGUSTUS CHAPLIN.

marriage bringing him the command of some capital, he entered into partnership with one Roberts, a coach-proprietor established at the "White Horse," Fetter Lane. But the partnership was dissolved at the expiration of twelve months, when Horne, making a bold stroke, purchased the "Golden Cross" of John Cross, who, having acquired a large fortune after many years in business there, was now retiring from it and entering upon a series of rash speculations which eventually ruined him and brought Thomas Cross, his son, down to poverty from the assured position of heir to that fortune, and thence to the dramatic reverse of soliciting employment as a coachman in the very yard his father once had owned.

Established thus at the "Golden Cross," William Horne further developed the very fine coaching business he had acquired, and added to it the yards at the "Cross Keys," Wood Street, and the "George and Blue Boar," Holborn, together with an office at 41, Regent Circus He soon had seven hundred horses in work, and was in the full tide of life and energy when he died in 1828, at the early age of forty-five. "His last journey," says the obituary notice of him, "was but a short distance—St. Margaret's churchyard, Westminster; and, as a man of talent, his remains were placed within a few feet of some of the greatest men of their age."

Benjamin Worthy Horne was thus only twenty-four years of age when the management

of this business fell to him. He soon had need
of all those fierce energies that were his, for, in
addition to a watchful eye upon the doings of
his rivals, he had the stress and turmoil of the
rebuilding of the " Golden Cross " to contend with.
To him, indeed, fell the singular experience of
having that central place of business rebuilt
twice in three years, and the second occasion on
another site. When it was first rebuilt, in 1830,
Trafalgar Square was not in existence, and the
inn was re-erected on the old spot at the rear
of Charles I.'s statue, exactly where the south-
eastern one of Landseer's four lions, guarding
the Nelson Column, now looks across towards
the Grand Hotel.

But no sooner was the place rebuilt than
the Metropolitan improvements in the meanwhile
decided upon brought about the clearance of the
site, and the present " Golden Cross " arose some
distance away. At this time fifty-six coaches
left that place daily, many of them bitterly competitive with those of other proprietors. Equally
with his father, Benjamin Horne was an extremely
keen business man, and eager to cut into any
paying route. He had stables at Barnet and
Finchley, to enable him to compete advantageously
on the northern and north-western roads with
Sherman, of the " Bull and Mouth," and with
others on those routes. As early as 1823, when
the " Tally-Ho ! " fast coach between London and
Birmingham was first put on the road by Mrs.
Ann Mountain, of the " Saracen's Head," Snow

Hull, to do the 109 miles in 11 hours, the success of her enterprise had roused the jealousy of William Horne, who speedily started the "Independent Tally-Ho!"—setting out an hour and a quarter earlier, in order to intercept the bookings of the original conveyance. Numerous other "Tally-Ho's!" were then established, and the racing between them on the London and Birmingham road grew fast and furious, much to the advantage of the slower coaches, whose bookings were wonderfully increased by timid passengers refusing to go any longer by these breakneck rivals.

Benjamin Worthy Horne had at one time seven mails: the old Chester and Holyhead; the Cambridge Auxiliary; the Gloucester and Cheltenham; the Dover Foreign Mail, the Norwich, through Newmarket; the Milford Haven; and the Worcester and Oxford, in addition to the Hastings, a two-horsed affair, afterwards transferred to the "Bolt-in-Tun" office in Fleet Street.

Urged on, perhaps, by the partial success of the competitive "Tally-Ho!" he started in 1834, in alliance with Robert Nelson of the "Belle Sauvage" and Jobson of the "Talbot" at Shrewsbury, the "Nimrod" London and Shrewsbury coach, to compete with that pioneer of long-distance day coaches the "Wonder," a highly successful venture established so early as 1825, by Sherman of the "Bull and Mouth," and Taylor of the "Lion" at Shrewsbury. The bitterness and bad blood thus stirred up were almost incredible. It is not to be supposed that men so spirited as Sherman and Isaac Taylor were content to idly see this late-comer enter the field their own enterprise had opened, and be allowed to cut up their profits; and so the following season witnessed the appearance of the "Stag," own sister to the "Wonder," and by the same proprietors, timed to run a little in advance of the "Nimrod," while the "Wonder" went slightly in the rear. Thus the hated rival was pretty well "nursed" all the way, and did not often succeed in securing a well-filled way-bill. The pace while this insane competition lasted was terrific, and the coachman of the "Nimrod" on the Wolverhampton and Shrewsbury stage was thrown off and killed. The coaches were originally fast, being timed at $11\frac{1}{2}$ miles an hour; but in the furious racing that took place, day after day, the whole three often arrived together at the journey's end, two hours before time. One shrinks from computing the pace an analysis of these figures would disclose. The fares by the "Wonder" and "Stag" were in the meanwhile reduced by one-third; and, partly in consequence of this "alarming sacrifice," and a great deal more, we may suppose, in consequence of travellers being afraid to travel by these reckless competitors, £1500 were lost by Sherman and his allies in twelve months. But at the end of that time they had the satisfaction of seeing the "Nimrod" withdrawn, when the fares were raised to their old level.

We are not told how much Horne and his friends lost in this onslaught upon Sherman's preserves, but it must have been a very considerable sum. Horne ran in opposition to many proprietors, and was powerful enough to wear down any competitors except the three or four men whose businesses ranked with his own for size. Those proprietors who agreed to work with rather than against him, were therefore the better advised. When putting a new coach on a route, his practice was to offer a share in the business to others accustomed to work along it. If they refused, and elected to oppose him, he became dangerous. He never allowed competition; and as he had the longer purse, generally beat his rivals. A strictly businesslike proprietor would accordingly always welcome Horne as a partner; but it generally happened that men who had for years past run coaches on certain roads fell unconsciously into the habit of thinking and acting as though they held a prescriptive right to the whole of the traffic along them, and not only refused to ally themselves with any one providing additional coaches, but endeavoured to shut him out altogether. Thus Horne, although ready to work with any proprietor, was in bitter opposition on many roads.

His was the Liverpool "Umpire," a day coach; and his, too, the "Bedford Times," so far as horsing it out of London was concerned. It was started about 1836, by Whitbread, the brewer, as a hobby, and ran from the "George

and Blue Boar." It is singular that it made
the third Bedford coach running daily from
that inn: Horne seems to have considered that
Bedford could not have too many coaches. The
others were the "Telegraph," twice a day—8 a.m.
and 2.15 p m.—and the "Royal Telegraph" at
9 a m   The "Times" started at 3 p m., and
went at $10\frac{1}{2}$ miles an hour, including stops. This
was a very smart and exclusive coach, built on
the lines of the private drag, and ran to that
monumental Bedford hotel, the "Swan." The
"Bedford Times" was further remarkable as one
of the last-surviving of the coaches. It was not
run off the road until 1848.

Horne prided himself on his drastic ways, and
was fond of recounting his master-strokes in
crushing out rivals. The particular coup on
which he loved to dwell was that of driving up
to an inn belonging to a middle-ground partner
of one of his enemies, and buying up all the
horses overnight, so that in the morning, when
his own coach bowled by, the rival concern was
brought to an ignominious standstill. This story,
if true, reflected no credit on either himself or
the other party to the transaction, who certainly
was liable to an action for breach of contract.
There is, however, no doubt at all that Horne
was the man to have gone to the extravagant
length of indemnifying the vendor—perhaps better
described as his accomplice—against any action-
at-law. He simply would not brook business
rivalry

THE "BEDFORD TIMES," ONE OF THE LAST COACHES TO RUN, LEAVING THE "SWAN HOTEL," BEDFORD.

He was a tall, lathy, irritable man, of eager face, quick, nervous speech, and rapid walk, with something of a military air in his alert, upright figure. The very antithesis of Chaplin, who was of short stature and possessed of a nature that nothing could ruffle, Horne must always expend

BENJAMIN WORTHY HORNE.

his energies on the minor details of his extensive business, and himself do work that would have been better delegated to subordinates. In the end this wore him out, and brought him to a comparatively early death. Up early, no day was long enough for him, and he economised time by taking no regular meal until evening. He was generally to be seen eating his lunch out of a paper bag as he swung furiously along the streets. "There's Horne," said one of those many who did not love him, "with the devil at his elbow, as usual!"

It was, perhaps, well for him that Chaplin, calm and level-headed, came and entered into discussion on the railway question at that critical time when the fortunes of coach-proprietors were to be saved or lost by a simple declaration of policy. The time was 1837, the occasion the approaching opening of the first section of the London and Birmingham Railway. Should they hold out against the new order of things, as Sherman was bent upon doing, or should they enter into that alliance with the railway for which the railway people themselves were diplomatically angling? Chaplin thought they should, and proposed an amalgamation of their two interests Horne was not so sure of railway success, and might have continued on his own way, but Chaplin, who was an old friend, urged his own views. "We shall lose £10,000 apiece if we don't work with them," he said, "and you won't like that, Benny, my boy." Eventually Horne agreed, and the firm of Chaplin & Horne was founded.

Dark rumours were current at the time that to this newly constituted firm a sum of several thousands of pounds was paid by the London and Birmingham directors as the price of their friendship, but, however that may be, the allied coach-proprietors agreed to withdraw their coaches from the Birmingham Road, and to throw the weight of their interest and influence on the side of the railway. In return, they were given the contract for the parcel agency of the line. Chaplin had perceived, as Baxendale had already done in the case of the goods traffic, that this agency would be very valuable, and to his far-seeing counsel Horne owed much.

Henry Horne, one of Benjamin Worthy Horne's nine brothers, became a partner with him in 1836, and was a member of this firm of Chaplin & Horne for many years. He survived his brother, and was at the head of affairs when the London and North-Western Railway took over the parcel business and the London receiving offices in 1874. Henry was the kindest-hearted of men, and old coaching-men down on their luck always found him a sure draw for a loan or a gift. Wise by dint of long experience, he laid down a golden rule that it was cheaper in the end to give £50 than to lend £100.

When the fierce old fighting days of the road were ended and the business of Chaplin & Horne was set afoot, the restless energies of Benjamin Worthy Horne found an outlet in the management of the goods business in connection with the railway, and he was constantly in and out at Euston and Camden. In those early days the London and North-Western Railway headquarters staff was managed on somewhat lax and primitive lines, and if a departmental manager thought he wanted a little holiday, he took it, without a word to any one. To a strict and keen business man like Horne these proceedings seemed particularly strange, and were often, doubtless, the source of much annoyance and waste of time. He had the unchallenged run of the offices, and was so used to finding the various managers away, on some pretext or another, that he would humorously assume their absence on all occasions. With his abrupt manner, he would burst boisterously into a room, and exclaim—

Ah! Manager Number One out—
Gone fishing, no doubt!

At the next office, whether the manager happened to be in or not, he would enter with the same assumption of his absence, and say—

Manager Number Two
Nothing to do—
Of course, gone fishing also!

To his especial aversion David Stevenson, the goods manager, whom he considered to have usurped many of his firm's rights and privileges, he would enter tragically with—

Aha! Manager Stevenson—
Gone about his private theatricals!

and fix the enraged Stevenson with the haughty stare common to the transpontine drama of the time. The sting of it lay in the fact that Stevenson belonged to an amateur dramatic society.

The goods department at Camden was taken

over by the London and North-Western Railway in Benjamin Worthy Horne's time, long before the general parcels and receiving-office branch was absorbed. The decision to terminate the contract was a source of much annoyance to him, on account of the reason given, which was that the business was not efficiently conducted. Although he was a man who in general had a horror of going to law, this stigma upon his business methods so stung him that he brought an action against the railway company for breach of contract, in order to vindicate his position. This was going to law for an idea, and as the company had a perfect right to terminate the contract, the action of course failed; but it was made abundantly evident that the business was efficiently carried on, and that the railway was only proposing to take it over because the time was ripe for such a development. His heavy costs, amounting to £1200, were afterwards very handsomely refunded to Mr. Horne by the railway.

It remains to say that although there was no keener or more ruthless man of business than Benjamin Worthy Horne, he was privately a considerate and kindly man, helpful and charitable to those less successful than himself.

He had a pretty estate at Highlands, Mereworth, and a town residence at 33, Russell Square. He died at the latter place, April 14th, 1870, aged sixty-six, leaving property valued at £250,000.

CHAPTER X

COACH-PROPRIETORS (*continued*)

EDWARD SHERMAN, who ranked next to Chaplin as the largest coach-proprietor in London, was in many respects unlike his brethren in the trade. He established himself at the " Bull and Mouth," St. Martin's-le-Grand, in 1823, in succession to Willans, and came direct from the Stock Exchange, where he had been a broker in alliance with Lewis Levy, a noted figure in those days of Turnpike Trusts It is perhaps scarcely necessary to add that Levy was a Jew. He was referred to by Lord Ravensworth in the course of a discussion in the House of Lords on Metropolitan Toll-gates in 1857 as "a gentleman of the Hebrew persuasion." Persuasion, indeed ! As well might you describe a born Englishman or Frenchman as born into those nationalities by personal choice and election. Levy was, of course, a Jew by birth, and had no choice in the matter. He was a farmer of turnpike-tolls to the extent of half a million sterling per annum, and a very wealthy man. Levy put Sherman into the coaching business, and he immediately began to make things extremely uncomfortable for the older proprietors, who had up to that time been content with going at eight or nine miles an hour.

When Colonel Hawker took coach from the "Bull and Mouth" in 1812, he found "the ruffians" there "a dissatisfied, grumbling set of fellows, and their turns-out of horses and harness beggarly." Such was the place under Willans' rule, but Sherman altered all that. He was anything but a horsy man, and it is therefore remarkable that he should have built up the very extensive business that the "Bull and Mouth" Yard did almost immediately become. He was the pioneer of fast long-distance day coaches, and was the proprietor, at the London end, of the "Shrewsbury Wonder," which, like all his coaches at that time, was a light yellow and black affair. How long he continued subservient to Levy may be a matter for conjecture, but when he rebuilt the "Bull and Mouth" Hotel, in 1830, he did so from the money of one of the three old and wealthy ladies whom he married in succession. The "Wonder" ran 158 miles in the day, as against the 122 miles to Bristol; but was shortly afterwards eclipsed by the Exeter "Telegraph," put on the road in 1826 in rivalry with Chaplin's "Quicksilver" Devonport Mail, by Mrs. Ann Nelson, of the "Bull," Whitechapel. In this Sherman had only a small share. Entirely his own venture was that supreme achievement, the "Manchester Telegraph" day coach, started in 1833 and running 186 miles in 18 hours, technically in the day by dint of starting at 5 o'clock in the morning and reaching Manchester at 11 p.m. The journey was at last shortened by one hour, when the pace, allowing twenty minutes for dinner at Derby, and stops for changing, worked out at just under twelve miles an hour. The Manchester "Telegraph" day coach must by no means be confounded with the old night coach of that name, which in 1821 started from the "Castle and Falcon" at 2.30 p.m., and arrived at the "Moseley Arms," Manchester, at 8 o'clock the next evening —29½ hours, not much more than six miles an hour.

The "Telegraph" day coach was built by Waude, and was able to safely perform its astonishingly quick journeys over what is in some places an extremely hilly road by the introduction of the flat springs that, from first being used on this coach, were known as "telegraph springs," a name they retain to this day. They set the fashion of low-hung coaches, which, in the lowering of the centre of gravity, retained their equilibrium at high rates of speed and when going round abrupt curves. Accidents, very numerous in those years, would have been even more frequent had it not been for this change.

The heated rivalry between Sherman's "Manchester Telegraph" and Chaplin's "Manchester Defiance"—continued for some years—was but one phase of a keen competition that raged all round the coaching world for the possession of the Manchester traffic. The "Swan with Two Necks" "Defiance" may be traced back to 1821, and even before that date, if necessary. In that year there was not a coach that went the distance in less than 27 hours, and in this first flight the

" Defiance " was included. It set out at 2.30 p m., and was at the " Bridgewater Arms," Manchester, at 5.30 the next afternoon. By 1823 it was accelerated by two and a half hours; in 1826 it had become the " Royal Defiance," in 24 hours. In succeeding years it continued to go at 6.30 and 6.15 p.m , and when the " Telegraph " was started the pace was screwed up to the same as that of the new-comer. An evening rival was the fast " Peveril of the Peak," running from the " Blossoms " inn, Lawrence Lane, Cheapside, while Robert Nelson, of the " Belle Sauvage," also had a fast night coach, the Manchester " Red Rover," at 7 p.m., a very lurid affair on which the guards wore red hats as well as red coats, and the horses red harness and collars as far as he horsed the coach out of London. This did not long remain in his hands. Sherman afterwards obtained it ; but Nelson, burning with professional zeal and no little personal pique, immediately put an entirely new coach on the same route to Cottonopolis. The announcement of the " Beehive," as it was called, is distinctly worth quoting, for it shows at once the keen rivalry between proprietors at this period and the excellent appointments of the later coaches :—

" NEW COACH FROM THE ' BEEHIVE ' COACH OFFICE

" Merchants, buyers, and the public in general, visiting London and Manchester, are respectfully informed that a new coach, called the 'Beehive,' built expressly, and fitted up with superior accommodation for comfort and safety to any coach in Europe, will leave 'La Belle Sauvage,' Ludgate Hill, London, at eight every morning, and arrive in Manchester the following morning, in time for the coaches leaving for Carlisle, Edinburgh, and Glasgow. Passengers travelling to the north will reach Carlisle the following morning, being only one night on the road. The above coach will leave the 'Beehive' Coach Office, Market Street, near the Exchange, Manchester, every evening at seven, and arrive in London the following afternoon at three. All small parcels sent by this conveyance will be delivered to the farthest part of London within two hours after the arrival of the coach. In order to insure safety and punctuality, with respectability, no large packages will be taken, or fish of any description carried by this conveyance. The inside of the coach is fitted up with spring cushions and a reading-lamp, lighted with wax, for the accommodation of those who wish to amuse themselves on the road. The inside backs and seats are also fitted up with hair cushions, rendering them more comfortable to passengers than anything hitherto brought out in the annals of coaching, and, to prevent frequent disputes respecting seats, every seat is numbered. Persons booking themselves at either of the above places will receive a card, with a number upon it, thereby doing away with the disagreeables that occur

daily in the old style. The route is through
Stockport, Macclesfield, Congleton, Newcastle,
Wolverhampton, Birmingham, Coventry, Dunchurch, Towcester, Stony Stratford, Brickhill,
Dunstable, and St. Albans, being the most level
line of country, avoiding the danger of the steep
hills through Derbyshire.

    " Performed by the public's obedient servants,

        " ROBERT NELSON, London ;

        " F. CLARE, Stony Stratford ;

        " ROBERT HADLEY & Co., Manchester."

Sherman's rebuilt " Bull and Mouth " inn,
or " Queen's Hotel," to give it its later name,
long remained a feature of St. Martin's-le-Grand,
many years after the last coach had been withdrawn ; and the old stables in Bull and Mouth
Street, which had not been included in the rebuilding of 1830, remained, a grim and grimy
landmark, put to use, as usually the case with
the old coach offices, as a receiving office for the
Goods Department of one of the great railways.
In later years the " Queen's Hotel " became the
property of that very thick-and-thin supporter
of and believer in the Tichborne Claimant, Mr.
Quartermaine East ; but the growth of Post Office
business made the site an exceedingly desirable
one for an extension, and in 1887 the house was
closed and demolished, and in the fulness of
time the gigantic block of buildings officially
known as " G.P.O. North " arose. Not only
were the sites of hotel and stables thus occupied,

but even Bull and Mouth Street was stopped up and built over. The still-existing Angel Street, close by, between "G.P O. North" and "G.P.O. West," marks where another coaching inn, the "Angel," once stood.

Robert Nelson, who entered so keenly into rivalry with Sherman over the Manchester business, was one of the three sons of Mrs. Ann Nelson, of the "Bull Inn," Whitechapel. Not the Bull "Hotel," for Mrs Nelson most resolutely set her face against that new-fangled word; and as an "inn" the house was known to the very last. An excellent inn it was—one of the very best It did not seem strange then, as undoubtedly it would now be, for so high-class a house to be situated in this quarter of London. Whitechapel of that time was vastly different from the disreputable place it is to-day; but the prime reason of so fine an inn as the "Bull" being situated here was that this was the starting-point of many routes into the eastern counties, and, just as railway hotels form a usual adjunct of railway termini, so did Mrs. Nelson possess an excellent hotel business in addition to the important and highly successful coaches that set out from her yard and stables.

The "Bull," Whitechapel, was sometimes—and with equal, if not better, exactness—known as the "Bull," Aldgate, for it was numbered 25 in Aldgate High Street. The relentless hand of "improvement" swept it away in 1868, but until that year it presented the picture of a typical old

English hostelry, and its coffee-room, resplendent with old polished mahogany fittings, its tables laid with silver, and the walls adorned with numerous specimens of those old coaching prints that are now so rare and prized so greatly by collectors, it wore no uncertain air of that solid and restful comfort the newer and bustling hotels of to-day, furnished and appointed with a distracting showiness, are incapable of giving. Everything at the " Bull " was solid and substantial, from the great heavy mahogany chairs that required the strength of a strong man to move, to the rich old English fare, and the full-bodied port its guests were sure of obtaining.

A peculiar feature of this fine establishment of Mrs. Nelson's was the room especially reserved for her coachmen and guards, where those worthies supped and dined off the best the house could provide, at something less than cost price. Mention has often been made of the exclusiveness of the commercial-rooms of old, but none of those strictly reserved haunts were so unapproachable as this coachmen's room at the " Bull." There they and the guards dined with as much circumstance as the coffee-room guests, drank wine with the appreciation of connoisseurs, and tipped the waiter as freely as any travellers down the road. A round dozen daily gathered round the table of this sanctum, joined sometimes by well-known amateurs of the road like Sir Henry Peyton and the Honourable Thomas Kenyon, but only as distinguished and quite exceptional guests. Once, indeed, Charles Dickens sat at this table. Perhaps he was contemplating a sequence of stories with some such title as "The Coachmen's Room"; but if so, he never fulfilled the intention. The chairman on this occasion, after sundry flattering remarks, as a tribute to the novelist's power of describing a coach journey, said, "Mr. Dickens, sir, we knows you knows wot's wot, but can you, sir, 'andle a vip?" There was no mock modesty about Dickens He acknowledged that he *could* describe a journey down the road (doubtless, if we have a correct mental image of the man, he acknowledged that little matter with a truculent suggestion in his manner that he would like to see the man who could do it as well), but he regretted that in the management of the "vip" he was not an expert.

Unlike commercial dinners, "shop" was not taboo round this hospitable mahogany, but formed the staple of the conversation. Indeed, these worthies could talk little else, and with the exception of sometimes shrewd and humorous sidelights on the towns and villages they passed on their daily drives, and criticisms of the local magnates whose parks and mansions they pointed out to the passengers on the way, were silent on all subjects save wheels, horses, and harness

The etiquette of this room was strict. The oldest coachman presided—never a guard, for they always ranked as juniors—and at the proper moment gave the loyal toast of the King or Queen. An exception to this rule of seniority was when

Mrs. Nelson's second son, Robert, who drove her Exeter "Defiance," was present, as occasionally he was. Following the practice of the House of Commons, whose members are never, within the House, referred to by their own names, but always as the representatives of their several constituencies, Mrs. Nelson's coachmen and guards here assembled were addressed as "Manchester," "Oxford," "Ipswich," "Devonport," and so forth.

When Mrs. Nelson retired from the active management of the business, her eldest son, John, became the moving spirit. It was in his time that railways came in and coaching went out, but he was equal to the occasion, and started a very successful line of omnibuses, the "Wellington," plying between Stratford, Whitechapel, the Bank, Oxford Street, Royal Oak, and Westbourne Grove. He died, a very wealthy man, in June, 1868, aged seventy-four

Thomas Fagg, of the "Bell and Crown," Holborn—an inn better known to later generations of Londoners as "Ridler's Hotel"—was a small proprietor, but he had in addition a very lucrative business as a coach-maker at Hartley Row, near Basingstoke. The "Louth" and "Lynn" mails, however, were partly his, and *Cary's Itinerary* for 1821 gives a list of twenty-six stage-coaches going from his door to all parts of the country As "Ridler's" the house was a very select "family hotel," but in this it only carried on the traditions of Fagg's time, when he had some most distinguished guests. Standing

midway between the West End and the City, the
" Bell and Crown " thus possessed certain advantages, and received much patronage both from
commercial magnates and Society people. Among
his patrons he numbered the " Iron Duke," for
whom he had an almost religious reverence, and
indeed proposed to change the name of his house
to the " Wellington," in honour of him ; only re-
considering the project when the Duke told him—
as he commonly did the many extravagant hero-
worshippers whose attentions were a daily nuisance
—not to be " a d——d fool." Fagg, however,
was no fool, but a very shrewd person indeed.
A coachman, applying to him for a place on one
of his coaches, was put through a strict examination as to his qualifications, when it appeared
that he was (according to his own account) not
only a first-rate and steady " artist," but had
never capsized a coach in the whole course of his
career—" he didn't know what a hupset meant."

" Oh ! go away," retorted the justly incensed
Fagg ; " you are no man for me. *My* coaches
are always upsetting, and with *your* want of
experience, how the devil should you know how to
get one on her legs again ? "

Mrs Mountain also had her own coach-factory.
She was no less energetic than that very lively
and masterful person, Mrs. Ann Nelson, but in a
smaller way of business. Sarah Ann Mountain's
house was that " Saracen's Head," Snow Hill,
immortalised by Dickens in *Nicholas Nickleby*.
She had succeeded to the business in 1818, on the

death of her husband, and instead of giving up, decided to carry on, aided by Peter, her son. Thirty coaches left her inn daily, among them the first of the Birmingham "Tally-Ho's," a fast day coach, established in 1823, and historically interesting as the prime cause of the furious racing that characterised the St. Albans and Coventry route to Birmingham from this date until 1838. Mrs. Mountain's coach-factory was at the rear of her premises on Snow Hill. There she built the conveyances used by herself and partners, charging them at the rather high rate of $3\frac{1}{2}d$. a mile for their use.

A number of smaller proprietors accounted, between them, for many other coaches. Robert Gray, once established at the " Belle Sauvage," left that place in 1807 and settled at the " Bolt-in-Tun," a house still standing in Fleet Street, and now known as the " Bolt-in-Tun " London and North-Western Railway Receiving Office. He sent out twenty-five coaches daily, almost exclusively down the southern and western roads, among them the Portsmouth and the Hastings mails, the latter a pair-horse concern.

William Gilbert, of the " Blossoms " inn, Laurence Lane, Cheapside, had also a pair-horse mail—the " Brighton "—the " Tantivy," Birmingham coach, and a fast night coach to Manchester, the " Peveril of the Peak." Seventeen other coaches left his yard

Joseph Hearn, proprietor of the " King's Arms," Snow Hill, was monarch among the slow-coaches, of which he had twenty-two. Among

them were the Bicester "Regulator," the Boston
"Perseverance," and the Leicester and Market
Harborough "Convenience"—names that do not
spell speed. Even his Aylesbury "Despatch"
was a slow affair, reaching that town in six hours,
at the rate of six and a half miles an hour.

Many great coach-proprietors were established
in the chief provincial towns. Bretherton, of
Liverpool, described by Chaplin as "an exceedingly opulent man," Wetherald, at Manchester,
Teather, of Carlisle, Waddell, at Birmingham,
are names that stand forth prominently. The
cross-country rivalry between these men was quite
as bitter as that which raged among the Londoners,
and, although with the lapse of time the exact
explanation of the following extraordinary epitaph
on a coach-proprietor of Bolton, Lancashire, cannot
be given, it is doubtless to be found in one of
these business feuds —

"Sacred to the Memory of Frederic Webb, Coach Proprietor,
of the firm of Webb, Houlden, & Co., of Bolton, who departed
this life the 9th December, 1825, aged 23 years. Not being
able to combat the malevolence of his enemies, who sought his
destruction, he was taken prematurely from an affectionate loving
wife and infant child, to deplore the loss of a good husband,
whose worth was unknown, and who died *an honest man*"

The inference intended to be drawn was
obviously that the others were not honest men;
but, honest or not, they are all gone to their
account, and the world has forgotten them and
their contentions. Only the stray historian of these
things comes upon their infrequent footmarks,
and wonders greatly at their elemental ferocity.

CHAPTER XI

THE AMATEURS

Those men ascend to lofty state,
And Phœbus' self do emulate,
Who drive the dusty roads along
Amid the plaudits of the throng.
When round the whirling wheels do go,
They all the joys of gods do know.
See the Olympian dust arise
That gives them kindred with the skies!

HORACE, Book 1., Ode I.

THUS Horace sings, in his Ode to Mæcenas; and the driving ambition observed by that old heathen, still to be noticed in these days, was a very marked feature of the road at any time between 1800 and 1848, when the railways had succeeded in disestablishing almost every coach, and the opportunities of the gentleman coachman were gone.

The amateur coachman was a creation of the nineteenth century. He was, for two very good reasons, unknown before that time. The first was that coachmanship had not yet become an art, and, still in the hands of mere drivers whose only recommendations were an ability to endure long hours on the box and a brutal efficiency in punishing the horses, had no chance of developing those refinements that characterised the Augustan age of coaching; the second reason was that the box-seat, although perhaps already beginning to be regarded as a place of distinction, was much more certainly a very painful eminence. It rested directly upon the front axle, and, being wholly innocent of springs, received and transmitted to the frame of any one who occupied it every shock the wheels encountered on the rough roads of that time.

Springs under the driving-box were unknown until about 1805, when they were introduced by John Warde, of Squerryes, the old Kentish squire who is generally known as the "Father of Fox-hunting." He was the first amateur coachman, and in pursuing that hobby found the driving-seats of the old coaches anything but comfortable. In resisting his arguments in favour of the intro-duction of springs, the coach-proprietors declared to a man that the coachmen would always be falling asleep if they were provided with com-fortable seats.

John Warde's driving exploits were chiefly carried out on the Oxford, Gloucester, and Bir-mingham roads. For years before coachmanship became a fashionable accomplishment, he had been accustomed to take the professional coachman's place on the "old Gloucester" stage, "six inside and sixteen out, with two tons of luggage"; or, relieving Jack Bailey and other incumbents of the bench on the old Birmingham and Shrewsbury "Prince of Wales," would drive the whole distance between London and Birmingham. He once drove this coach from London to Oxford against the

"Worcester Old Fly" for a wager, and won it, although his coach went the Benson road, four miles longer than the route his opponent had to travel.

Warde's driving was by no means in the later style, and he probably would have been very much out of his element with the smart galloping teams of the Golden Age. He was, however, of those who were fit to be trusted with a heavy load behind weak horses and on bad roads. There was a peculiarity about him as regarded the driving of his own horses which the history of the road, it was said, could not parallel. Let the journey be in length what it might, he never took the horses out of his private coach, giving them only now and then a little hay and a mouthful of water at a roadside public-house. When he resided in Northamptonshire, sixty-three miles from London, the journey was always accomplished by his team "at a pull," as he called it. The pace, as may be supposed, was not quick. John Warde was one of the founders of the B.D.C., or Benson Driving Club, in 1807.

Amateur coaching, as a fashionable amusement, took its rise on the Brighton Road. Looked upon with contempt by stalwart and bluff Warde and his kind, it nevertheless grew and flourished in the hands of the Barrymores and their contemporaries, Sir John Lade and Colonel Mellish; and in the early years of the nineteenth century the education of no gay young blood was complete until he had acquired the art of driving four-in-hand, in addition to the already fashionable and highly dashing sport of driving the light whiskies, the high-perched curricles, and the toppling tilburies that then gave a fearful joy to the newly-fledged whip There was not too much physical exertion, endurance, or skill required on the road to Brighton, which was only fifty-two miles in length, and already possessed a better surface than most roads out of London; and, moreover, it was a road peopled from beginning to end with fashionables, before whom the gentleman-coachman could display his prowess. It was then pretty generally recognised that coach-driving was a poor sport if the ease and grace of the performer could not be displayed before a large and fashionable audience. That, it will be conceded, was not altogether a worthy attitude.

Many of these brilliant amateurs of the road ran an essentially identical career of viciousness and mad extravagance; and not a few of them wasted themselves and their substance in the very shady pursuits that then characterised the "man about town." Those who are curious about such things may find them fully set forth in Pierce Egan's *Life in London* and its grim sequel, the *Finish*. The endings of the Toms and Jerrys of that Corinthian age were generally sordid and pitiful.

The truth is that the sporting world was then, as it always has been and always will be, thronged with the toadies who were ever ready to fool a moneyed youngster to the top of his bent. He

FOUR-IN-HAND.

After G. H. Laporte.

must vie with the richer and the more experienced, though he ruin himself in the doing of it, and bring his ancestral acres to the hammer, in the manner of a Mytton or a Mellish. The only satisfaction these reckless sportsmen obtained, beyond the immediate gratification of their tastes, was the eulogy of the sporting scribes, who discussed their style upon the box-seat with as much gravity as would befit some question of empire. Excepting "Nimrod" and "Viator Junior," whose essays on sport in general, and coaching in particular, were sound and honest criticism, these writers were venal and beneath contempt.

A "real gentleman," according to the ideas of these parasites, was one who flung away his money broadcast in tips. Many foolish fellows, foolish in thinking the good opinions of these gentry worth having, spent their substance in this way. Of this kind was the amateur whip described by a writer in the *Sporting Magazine* in 1831. This aspirant for the goodwill of the stable-helpers and their sort sat beside the professional coachman on the Poole Mail starting from Piccadilly, and when the reins were handed to him proclaimed his gentility by the distribution of shillings among the horsekeepers. First "Nasty Bob," the ostler, got a shilling for talking about the leaders' "haction"; then "Greedy Dick," the boots, had one also for handing him the "vip"; and then came "Sneaking Will," the cad and coach-caller, to say something civil to

the " gemman "; and even the neighbouring
waterman was seduced from his hackney-coaches
to try the persuasive powers of his eloquence.
Four shillings and sixpence this " real gentleman " distributed at Hatchett's door, and left
the capital with the best wishes of the donees
for his safe return. His generosity was not
allowed a long respite, for at " that vile hole
Brentford," a slowly manœuvring waggoner
blocked the way; and finding that he could by
no other means be induced to allow the mail to
pass, our amateur descended from the box, and,
slyly placing a shilling in the waggoner's hands,
said in a loud voice, " I don't stand any nonsense,
you know, so now take your waggon out of the
way. This forcible and intelligible appeal, so
properly accompanied, was perfectly irresistible :
the waggon was drawn to the roadside, and the
mail proceeded

Very few of these amateurs have been considered worthy of biographical treatment, but
among them Sir St. Vincent Cotton is one. Let
us just see what the outline of his life was :—
" Cotton, Sir St. Vincent, 6th Baronet, son of
Admiral Sir Charles Cotton. Born at Madingley
Hall, Cambs., October 6th, 1801; succeeded,
February 24th, 1812 ; educated at Westminster
and Christ Church, Oxford. Cornet 10th Light
Dragoons, May 13th, 1827 ; Lieutenant, December
13th, 1827, to November 19th, 1830, when placed
on half-pay. Distinguished himself in the hunting,
skating, racing, and pugilistic world. Played in

Marylebone Cricket Matches, 1830-35   A great
player at hazard.   Dissipated all his property.
Drove the 'Age' coach from Brighton to London
and back for some years from 1836.   Died at 5,
Hyde Park Terrace, January 25th, 1863."

It is possible to largely supplement this
skeleton biography from the *Sporting Magazine*
and other sources.   "The Cottons of Madingley
and Landwade," said that classic authority, "are
no 'soft goods' of recent manufacture, but have
held high rank among the gentry of Cambridge-
shire since the reign of Edward I.   Sir John
Cotton, the first baronet of the family, was advanced to that honour in 1641, by Charles I.,
to whose cause he was firmly attached.   Sir
St. Vincent used to ride in the first flight with
the crack men of Leicestershire, mounted on his
favourite mare, 'Lark.'   The honourable baronet
has, however, left both the Army and the Chase
to devote himself exclusively to the public service
on the 'Road,' where he performs the duties of a
coachman very much to his own pleasure, and the
great satisfaction of all His Majesty's lieges who
travel by the Brighton 'Age'; and we are of
opinion that an English baronet is much better
employed in driving a coach than in endeavouring
—like a certain mole-eyed wiseacre of the West,
who also displays the Red Hand on his scutcheon
—to saw off the branch that he is sitting on.

"We believe that the late Mr. H. Stevenson,
who drove the 'Age' a few years ago, was one
of the first gentleman-whips who took a *bob* and

returned a *bow*—*i.e.*, if you popped a shilling into his hand at the end of a stage, he ducked his head and said, 'Thank you.' The example thus set has been followed by the Baronet, who receives a 'hog' as courteously as his predecessor. When a noble Marquis, now in the enjoyment of an hereditary dukedom, drove the 'Criterion,' and afterwards the 'Wonder,' also on the Brighton Road, he did not take 'civility money,' we believe, but did the thing for pure love.

> "By different means men strive for fame,
> And seek to gain a sporting name.
> Some like to ride a steeple-chase;
> Others at Melton go the pace,
> Where honour chief on him awaits
> Who best takes brooks, and rails, and gates,
> Or tops the lofty 'bullfinch' best,
> Where man and horse may build a nest;
> Who crams at everything his steed—
> And clears it too—and keeps the lead
> Some on the 'Turf' their pleasure take,
> Where knowing 'Legs' oft bite 'the Cake';
> Others the 'Road' prefer; and drest
> Like 'reg'lar' coachmen in their best,
> Handle the ribbons and the whip,
> And answer 'All right!' with 'yah hip!'
> At steady pace off go the tits,
> Elate the Sporting Dragsman sits;
> No peer nor plebeian in the land
> With greater skill drives four-in-hand.'

Cotton, known to the plebeian professionals of the Brighton Road as "the Baronet," and to his familiars as "Vinny," was so hard hit by his disastrous gambling that he owned and drove the Brighton "Age" for a living. Let us do him the

SIR ST. VINCENT COTTON

justice to add that he did not attempt to disguise the fact, and that he took his misfortunes bravely, like a sportsman. Reduced, as a consequence of his own folly, from an income of £5000 a year to nothing, " I drive for a livelihood," he said to a friend : " Jones, Worcester, and Stevenson have their liveried servants behind, who pack the baggage and take all short fares and pocket all the fees. That's all very well for them. I do all myself, and the more civil 1 am (particularly to the old ladies) the larger fees I get " He, indeed, made £300 a year out of this coach, and got his sport for nothing.

The " Jones " of whom he spoke was Charles Tyrwhitt Jones, of whom, being just an amateur with no eccentricities, we know little. Of Harry Stevenson, one of the most distinguished and accomplished among amateurs of the road, we know a good deal, although even of his short life full particulars have never been secured. He made his first appearance on the Brighton Road in August 1827, as part-proprietor of the " Coronet," and even then his name seems to have been one to conjure with, for it was for painting it on a coach of which he was not one of the licensees that Cripps was fined in November of that year. Stevenson was then but little more than twenty-three years of age. He had gone from Eton to Cambridge, and during his exceptionally short career was always known by the fraternity of the road as " the Cambridge graduate." Although so little is known of him, sufficient has come down to us to place him on a higher pedestal than that of the majority of the gentlemen amateurs. He was not only a supreme artist with the ribbons, "whose passion for the *bench*," as "Nimrod" says, "exceeded all other worldly ambitions," but he was also a supremely good fellow, in a broader and better significance of that misused term than generally implied. That he was one of the spendthrifts who had run through their money before taking to the road as a professional would appear to be a baseless statement, invented perhaps to account for that higher form of sportsmanship which entirely transcended that of the general ruck of "sportsmen," by inducing him to drive his coach, as an ordinary professional would, day by day, instead of when fine weather and the inclination of the moment served. A good professional he made, for he did by no means forget his birth and education when on the box, and was singularly refined and courteous. His second, and famous, coach was the "Age," put on the Brighton Road in 1828. This celebrated coach eclipsed all the others of that time, from the mere point of view of elegance and comfort. On a road like that to Brighton there was not, of course, the chance to rival such flyers as the Devonport "Quicksilver" and other long-distance cracks, but in every circumstance of its equipment it was pre-eminent. It was not for nothing that Stevenson loved the road. His ambition was to be first on it, and he succeeded. The "Age" was built and finished, horsed and found in every way without regard to cost. In a time when brass-mounted harness was your only wear, his was silver-plated. The horse-cloths, too, exhibited this unusual elegance, for they were edged with deep silver lace and gold thread, and embroidered in each corner with a royal crown and a sprig of laurel in coloured silks and silver. These cloths were, many years afterwards, presented to the Brighton Museum by Mr. Thomas Ward Capps, a later proprietor of the "Age," and they are still to be seen there.

This was not by any means the sum of Stevenson's improvements The usual guard he replaced by a liveried servant, whom he caused to attend upon the passengers, when the coach changed horses, with silver sandwich-box and offers of sherry of a kind that appealed even to the jaded palates of connoisseurs. Stevenson was as excellent a whip as he was a good-hearted gentleman. "I am not aware," wrote "Viator Junior," "if, to quote a vulgar saying, he was 'born with a silver spoon in his mouth,' but I certainly think he must have been brought into the world with a whip and reins in his hand, for in point of ease and elegance of execution as a light coachman he beats nineteen out of twenty of the regular working dragsmen into fits, and as an amateur is only to be approached by two or three of the chosen few "

Of course, coaching on these luxurious terms resulted in a staggering loss, and could not long have continued, but even those short possibilities were ended by the early death of Stevenson. The cause of the attack of brain-fever that ended his career early in 1830 is imperfectly known, and is merely said to have been "an accident." The last scene was pathetic beyond the ordinary. Exhausted at the end of delirium, the bandages that had held his arms were removed, when, feebly raising himself up in bed and assuming as well as he was able his old habitual attitude upon the box, he exclaimed, as if with the reins in his hand, and to his favourite servant, who usually stood at his leaders' heads, "Let them go, George; I've got 'em!" and so sank down, dying, upon his pillow, in the happy delusion of being once more upon the road.

Mr. Harry Foker and others of the "young Oxonians" or "young Cantabs" with more taste for driving four-in-hand than knowledge of that very difficult art, were frequent aspirants for the ribbons, and as they were generally flush of money and free with it, they often tasted the delights of tooling a coach along the highway. Professional coachmen on the Oxford and Cambridge roads reaped a bounteous crop of half-guineas by resigning the reins into these hands, but equally plentiful was the harvest of bruises and shocks gathered by the passengers as a result of their reckless or unskilled driving. These chartered libertines of the road are mentioned with horror by travellers in the first half of the nineteenth century, who have pictured for us four horses galloping at the incredible speed of

THE CONSEQUENCE OF BEING DROVE BY A GENTLEMAN.

After H. Alken.

twenty miles an hour, and the coaches rocking
violently, while the "outsides" hold on like
firemen, behind some uncertificated young cub
from Oxford or Cambridge, or, anticipating the
final cataclysm, drop off behind or dive into
the hedges.

Even more than the passengers, coach-proprietors dreaded amateur coachmen, and very
properly dismissed those professionals whom they
caught allowing the reins out of their charge.
They had cause for this dread, for not only
was the act of allowing amateurs to drive itself
an illegal one, entailing penalties, but it often
resulted in accidents, bringing in their train
very heavy compensation claims. Juries invari-
ably satisfied themselves as to whether a professional or an amateur was driving at the time
when an accident occurred, and assessed damages
accordingly.

Sir St. Vincent Cotton was the cause of a
serious accident that happened to the "Star of
Cambridge." Springing the horses over a favour-
able stretch of galloping-ground, he went at such
a reckless pace that Jo Walton, the professional
coachman, seized hold of the reins In doing
so the coach was overturned, and the passengers
severely injured. A jockey named Calloway
had his leg broken, and, with others, brought
an action for damages. The affair cost Robert
Nelson and his partners nearly two thousand
pounds.

A good amateur coachman was, as a general

rule, like an accomplished violinist, only to be produced by long training. Caught young and properly schooled, he might become an elegant as well as a thorough whip; but the late-comer rarely attained both grace and complete mastery. "He who would master this most fascinating science of coachmanship," says Dashwood, in the *New Sporting Magazine*, "must begin early, under good tuition. He must work constantly on all kinds of coaches, and, thereby accustoming himself to every description of team to be met with, no matter how difficult or unpleasant, will ere long acquire a practical knowledge on that all-important point, the art of putting horses well together." He then proceeds to sigh for one hour of "old Bill Williams," of the "Oxford Defiance," who, as a schoolmaster of gentlemen-aspirants to coaching honours was, in his time, unequalled. He was supposed to have turned out more efficient coachmen than all the rest of his brethren put together. "Never by any chance—confound him !—would he allow an error or ungraceful act to escape unnoticed, and I have often got off his box so annoyed at his merciless reproofs and lectures that I vowed no power on earth should make me touch another rein for him. The first morning, in particular, that I was with him I shall never forget. In spite of all my remonstrances, nothing would satisfy him but I must take the reins from the door of the very office, at the 'Belle Sauvage,' he himself getting up behind, in order, as he said, not to 'fluster

the young 'un.' By great good luck we got
pretty well into the street, and, without anything
worth telling, for some way past Temple Bar;
but, as my evil star would have it, the narrow
part of the Strand was uncommonly full, and
having rather an awkward team, and being moreover in a pretty particular stew, we had more
than one squeak at sundry posts, drays, etc., etc.
Still, not a word was uttered by the artist, though
by this time he had scrambled in front, till, after
a devil of a mistake in turning into the Hay-
market, he touched my arm very civilly, with a
' Pull up, if you please, sir, by that empty coal-
cart.' I did so—at least, as well as I could—and
found, to my utter horror, that it was for the
purpose of his requesting the grinning blacka-
moors that belonged to it *to lend him some six
or seven of their sacks, to take the dray home*;
'for,' said he, ' I am sure the gentleman won't
take it up to the Gloucester Coffee House *a
coach*.' "

CHAPTER XII

END OF THE COACHING AGE

"This is the patent age of inventions."—BYRON

IN 1789, Dr. Erasmus Darwin, of Shrewsbury, in writing his poem, the *Loves of the Plants*, penned a most remarkably accurate prophecy, comparable with Mother Shipton's earlier "carriages without horses shall go." He wrote :—

> Soon shall thy arm, unconquered steam, afar
> Drag the slow barge, or urge the rapid car ;
> Or on wide waving wings expanded bear
> The flying chariot through the realms of air
> Fair crews, triumphant, smiling from above,
> Shall wave their fluttering kerchiefs as they move,
> Or warrior bands alarm the gaping crowd,
> And armies shrink beneath the rushing cloud

The first part of this prophecy was fulfilled in the period between 1823 and 1833, when steam-carriages—the motor-cars of that age—had a brief popularity.

Before railways successfully assailed the coaches, horsed vehicles had faced the inventions of a number of ingenious persons who wrestled with that problem of steam traction on common roads which had attracted Murdock in 1781. Trevithick took it up in 1800, and others followed ; but it was not until 1823 that the subject began greatly to interest engineers. At that period, however, Hancock, Ogle, Church, Gurney, Summers, Squire, Maceroni, Hills and Scott-Russell plunged into that troubled sea of invention Chief among these, from the standpoint of results achieved, were Mr. (afterwards Sir) Goldsworthy Gurney, Walter Hancock, and Colonel Maceroni. Gurney as early as 1827 had patented and tried a steam-carriage on the road. The boiler, it was explained for the benefit of nervous people, was perfectly safe. Even if it were to burst, being " constructed on philosophical principles," no one could be hurt. On July 28th, 1829, he ran one of his inventions on the Bath Road. This was what he termed a " steam-tractor," used as an engine to draw an ordinary barouche. Unfortunately for Gurney, he and his party reached Melksham on the annual fair-day, and a hostile crowd of rustics not only surrounded the steam-carriage, shouting " Down with machinery ! " but stoned the engine, the carriage, and Gurney and his friends, with such effect that the machinery was disabled and several of the party very seriously injured.

But he evidently travelled the kingdom pretty extensively with his machines, for he agreed with one Mr. Hanning to grant him the right of working them on a royalty on the West of England roads, and entered into similar arrangements on the routes between London, Manchester, and Liverpool, London and Brighton, London and Southampton, and London, Birmingham, and Holyhead. Their price was agreed upon—to be

hired at 6d. a mile, or to be sold by Gurney at
£1000 each   During four months at the beginning
of 1831, Sir Charles Dance, who had bought some
of the carriages, established a steam service on the
road between Cheltenham and Gloucester   Three
double journeys a day were made, 396 regular
trips in all, covering 3644 miles, and conveying
2666 passengers, who paid £202 1s. 6d. in fares.
The enterprise was just beginning to show a profit
when the local Trusts secured an Act under which
they raised the tolls against steam-carriages to a
prohibitive height, and even went so far as to
obstruct the roads with loose gravel and stones,
with the result that the axle of one machine was
broken.

In June 1831 the "philosophical" boiler of
one of Gurney's steam-carriages, warranted not
to burst disastrously, exploded at Glasgow, and
seriously injured two boys.   Tom Hood wrote :—

> Instead of *journeys,* people now
>   May go upon a *Gurney,*
> With steam to do the horses' work
>   By power of attorney ,
>
> Tho' with a load it may explode,
>   And you may all be undone ,
> And find you're going up to Heaven,
>   Instead of up to London.

Yet a Select Committee of the House of Commons,
which had been appointed to consider the question
of steam-carriages, reported, four months later,
that such carriages could be propelled at an
average rate of ten miles an hour; that they

would become a cheaper and speedier mode of
conveyance than carriages drawn by horses, and
that they were perfectly safe (¹).

Between 1832 and 1838 there were no fewer
than seven important Steam-Carriage Companies
in existence, and probably, had it not been for
the hostility of Turnpike Trusts all over the
country, the roads would have been peopled with
mechanically-propelled vehicles. But tolls were
raised to such a height against the new-fangled
inventions that it became commercially impossible
to run them. Between Liverpool and Prescot
the 4s. toll for a coach became £2 8s. for a
steam-carriage; between Ashburton and Totnes
the 3s. impost became £2.

Evidently, from a coloured print published in
1833, Goldsworthy Gurney projected a London
and Bath service, but the turnpike authorities
crushed that also. An inscription under the
original print obligingly tells us all about this
type of Gurney's carriages .—

"The Guide or Engineer is seated in front,
having a lever rod from the two guide-wheels, to
turn and direct the Carriage, and another at his
right hand, connecting with the main Steam Pipe,
by which he regulates the motion of the Vehicle—
the hind part of the Coach contains the machinery
for producing the Steam, on a novel and secure
principle, which is conveyed by Pipes to the
Cylinders beneath, and by its action on the hind
wheels sets the Carriage in motion. The Tank,
which contains about 60 Gallons of water, is

placed under the body of the Coach, and is its full length and breadth. The Chimneys are fixed on the top of the hind boot, and, as Coke is used for fuel, there will be no smoke, while any hot or rarified air produced will be dispelled by the action of the Vehicle. At different stations on a journey, the Coach receives fresh supplies of fuel and water. The full length of the Carriage is from 15 to 20 feet, and its weight about 2 tons. The rate of travelling is intended to be from 8 to 10 miles per hour. The present Steam Carriage carries 6 inside and 12 outside Passengers. The front Boot contains the Luggage. It has been constructed by Mr. Goldsworthy Gurney, the Inventor and Patentee."

Gurney was held, by a Parliamentary Committee, to be "foremost for practical utility"; but that statement was owing, there is little doubt, to the influence of his many friends in Parliament. Hancock's steam-carriages were at least as efficient—but then he had no such influential supporters. Gurney claimed to have lost £36,000 directly in his experiments, and a much larger sum indirectly, through the excessive tolls imposed, and brought his grievances before Parliament. A Committee recommended a grant of £16,000 to him, as the first to successfully apply steam-carriages to use on public roads.

In 1824 Walter Hancock was experimenting on similar lines, but it was not until 1828 that a proposal was made to run a service of steam-carriages between London and Brighton, and not

GOLDSWORTHY GURNEY'S LONDON AND BATH STEAM-CARRIAGE, 1833.

Alec G. Morton.

until November 1832 that his "Infant" actually
made the attempt. It had already, at the
beginning of 1831, plied for public service as an
omnibus between Stratford and London, and now
was to essay those 52 miles between London and
the sea.

It performed the double journey, but, owing
to lack of fuel on the way, not in anything like
record time, although it is said in places to have
attained a speed of 13 miles an hour

In 1833 Hancock started a steam omnibus
between Paddington and the City, and by 1836
had three Between them, they conveyed no
fewer than 12,761 passengers. They were named
the "Era," "Autopsy," and "Automaton." Why
the middle one should have been named in a
manner so suggestive of accidents and post-
mortem examinations is not clear. But indeed,
the names of old-time and modern motor-cars
and their inventors, strange to say, generally
have been, and are now, sometimes singularly
unfortunate. Thus, in 1824, a Scotch inventor
of Leith produced a steam-carriage His name
was Burstall! Among recent motor-cars are the
"Mors" and the "Hurtu."

In October 1833 Hancock ran the "Autopsy"
to Brighton in 8½ hours (including three hours
in stops on the way), and later had successful
trips to Marlborough and back and Birmingham
and back. These performances were considered
so promising that a "London and Birmingham Steam-Coach Company" was formed, and

more steam-coaches ordered to be built. Fares between London and Birmingham were not to exceed £1 each, inside, and 10s. out Hancock, a thorough believer in his invention and its capacity for solving the road-problems of the time, offered to carry the mails at 20 miles an hour; but the Post Office declined. Railways had, in fact, just succeeded in attracting attention, and were so strongly supported by capitalists that steam-carriages suffered neglect, and their inventors were utterly discouraged. Bright hopes and prospects gradually faded away, and by 1838 the railways held the field, undisputed.

Railways themselves were at first ridiculed, and suffered from the necessity of obtaining Parliamentary sanction at a period when the landowning interests and public opinion were decidedly hostile. Even when their construction was authorised, every one ridiculed the railways, and called those people fools who had invested their money in them. To be a railway shareholder was at that time, to the majority of people, proof positive of insanity, while engineers and directors were regarded as curious compounds of fools and rogues. Any time between 1833 and 1837, the coachmen on the Great North Road would point out to their box-seat passengers the works of the London and Birmingham in progress beside that highway, and distinctly visible all the way between Potter's Bar and Hatfield and at various other points. "Going to run us off the road, *they say*," a coachman would remark, jerking his elbow and nodding his head towards the place where hundreds of navvies were delving in a cutting or tipping an embankment. Then, squirting a stream of saliva from between his front teeth, in the practised manner assiduously cultivated by admiring amateurs, he would lapse into a contemplative silence, quite undisturbed by any suspicion that the railway really would run the coaches off. The passengers by coach were nearly all of the same mind. Some thought the railways would be useful in carrying goods, but declined to believe that they or any one else would ever travel by them; and a large proportion of the railway directors and proprietors shared the same opinion, being quite convinced that railways would convey heavy articles and general merchandise, and that coaches would continue to run as of old. Lovers of the road, coachmen and passengers alike, called the engines "tea-kettles," protested that coaching had nothing to fear, and wished their heads might never ache until railroads came into fashion. They declared they would never—no, *never*—go by the railroad, but at length, when some urgent occasion arose, demanding speed, they trusted their precious persons in a railway train, and, to their surprise, found it "not so bad after all." The next occasion, such a person going to town would shrink as he encountered the "Swallow" coach, by which he had always travelled, and would feel guilty as he shook his head to the coachman's "Coming by me this morning, sir?" Why?

Because he had made up his mind to go by train, and so save something in time and pocket. This time our traveller rather liked it; and thus the "Swallow," and many another coach not already withdrawn, was doomed.

Let us follow the career of such a coach, to its last days.

Deprived of its best passengers, the exchequer of our typical "Swallow" began to decline. The stalwarts, whose love for the road was superior to economy of time and money, were faithful, but they were not numerous enough, and did not travel sufficiently often, for the old style of that fast post-coach to be maintained, so it was reduced from four horses to three. In coaching parlance, it ran "pickaxe," or "unicorn." No connoisseur in coaching matters would condescend to travel as a regular thing by a three-horse coach, and so those supporters were alienated, and, against their will, driven to the railway; and the "Swallow," badly winged, carried only frightened old women who looked upon steam-engines as wild beasts. As they died away, no one took their places, and the old concern became a pair-horse coach. The coachman had seen the change coming, and declared he would never be brought so low as to drive two horses. He had said the same thing when it was proposed to have three. "Drive unicorn!" he had said: "never!" But he did, and he drove pair-horse as well, when the time came. It was better to do so than to lose his place and face starvation.

By this time the iron had entered the soul of
our poor old friend, and had rusted there. He
who had been so smart and gay, with song and
joke and always good-humoured, suffered, like the
coach, a strange and pitiful metamorphosis. The
stringency of the times had thinned the establishment, and in the absence of ostlers and stablemen
he put in the horses himself, badly groomed, and
the harness dirty. No one washed or cleaned the
coach, and it ran with the mud and dirt of many
journeys encrusted on its sides. His coat grew
seedy, his gloves soiled. Instead of the silver-
mounted whip he had wielded for years, he used
one of common make. The old one, he said, had
gone to be repaired, but somehow or another the
job was never completed. At any rate, no one
ever saw the old whip again. At the same time his
smart white hat disappeared and was replaced by
a black one: observant people, however, perceived
that it was the identical hat, disguised by process
of dyeing. He could sink no deeper, you think.
But he could, and did. Even the short journey to
which the old "Swallow" had in course of time
been reduced by railway extensions came at last to
an end ; and then he drove the "Railway Bus"
to and from the station, with one horse. His
temper, once so high-mettled, had by now grown
uncertain. He was like an April day—stormy,
dull, gloomy, and with fitful gleams of sunshine,
all in turn. No one knew quite how to take
him, and every one at last left him very much
to himself. He was never a favourite with the

"commercial gentlemen," who were now his most frequent passengers, for he had always in the old days looked down upon any one under the rank of a county gentleman, and could by no means rid himself of that ancient attitude of mind Indeed, he lived in the past, and when he could be induced to talk at all, would generally be reminiscent of better days. Commencing with the unvaried formula, "I've seen the time when.   ." he would then proceed to draw comparisons, much to the disadvantage of present time and present company. He was then absurdly surprised when acquaintance, tired of these tactless speeches, avoided him Not so quick in his movements as of yore, and always impatient of dictation, he resented the bluff impatience of a " commercial " one morning, and when that " ambassador of commerce " desired him to " look alive there, now, with those boxes," flung the boxes themselves on the ground, and told that astonished traveller to " go and be damned ! " Unfortunately, although the traveller would have overlooked the insolence, he could not afford to disregard the loss of his samples, which happened to be china, and were all smashed.   He reported the occurrence to the hotel-proprietor, who, being a compassionate man, explained, as he instantly dismissed the offender, that he was very sorry, but he could not afford to keep so violent a man in his employ.

After this dramatic incident the ex-coachman hung about the station, and obtained a few, a very few, odd jobs as porter, until one day a gentleman

alighting from a train saw him With surprise
and sorrow in his eyes he recognised the once
smart coachman, who, years before, had tutored
him in driving. "Good God!" he exclaimed:
" is it you ? " The old man burst into tears.

He ended more happily than, but for this
chance, would have been the case, for the Squire
took him into his service, and there he remained
until he followed his generation to the Beyond.

The opening of the London and Birmingham
Railway in September 1838 did not suddenly
bring the Coaching Age to a close. Many
routes remained for years afterwards practically
unassailed, and even on the road to Birmingham
some coach-proprietors struggled with great spirit
against the direct competition of the railway.
At the close of 1838 a newspaper is found saying:
" A few months ago no fewer than twenty-two
coaches left Birmingham daily for London. Since
the opening of the railway that number has been
reduced to four, and it is expected that these will
be discontinued, although the fares by coach are
only 20s. inside and 10s. outside, whilst the fares
for corresponding places on the railroad are 30s.
and 20s."

Prominent among those men who declined to
give up without a struggle was Sherman, of the
" Bull and Mouth," whose coaches had run to
Birmingham, Manchester, and other places on
the north-western road. For two years he maintained the unequal contest, and only relinquished
it when he had lost seven thousand pounds and

found his coaches running empty. Before finally beaten, he had even gone the length of re-establishing some coaches originally withdrawn in 1836, on the opening of the Grand Junction Railway. The reasons for this were many. The train-service in those early days was very poor, and engine-power insufficient, so that heavy loads, rain-showers that made the rails slippery, and the innumerable minor accidents always happening to the engines themselves, made travelling by railway not only uncertain, but, in not a few instances, even slower than by coach. Railway officials, too, were insolent to an incredible degree. Only when one has read the "Letters to the Editor" in contemporary journals can we have any idea of that insolence. The public complained that, having run the coaches off and secured a monopoly, the officials, finding themselves masters of the situation, behaved accordingly like masters, and not like the servants of the public they really were, or should have been. Newspaper comments dotted the i's and crossed the t's, and generally emphasised and embroidered these grievances. It is not, then, to be wondered at that a regret for "the good old times" found expression, or that coaches reappeared for a while. Many coach-proprietors were deceived by this partly indignant, partly sentimental attitude, and when they had committed themselves to a revival did not find the support which, from the newspaper outcry, they might reasonably have expected. Thus early do

THE LAST JOURNEY DOWN THE ROAD.

After J. H. Agase.

we find that gigantic evil of modern times—
irresponsible and misleading newspaper talk—
directly to blame for losses and disappointments
to those foolish enough to pay heed to it.

Sherman's country partners were not so rash
or so obstinate as he, and some of the coaches
he personally would have continued had been
withdrawn early in the railway advance. Among
those was the Manchester "Red Rover"; but
when the popular indignation against railway
delays and official insolence was thus exploited
by the newspapers, Sherman was enabled to again
secure the co-operation of his allies, and to put
that coach on the road once more. The decision
to do so was announced in a striking handbill :—

"The Red Rover re-established
throughout to Manchester.

Bull and Mouth Inn and Queen's Hotel.

It is with much satisfaction that the Proprietors of the Red Rover Coach are enabled to
announce its

Re-establishment

as a direct conveyance THROUGHOUT, BETWEEN
London and Manchester, and that the arrangements will be the same as those which before
obtained for it such entire and general approval
In this effort the Proprietors anxiously hope that
the public will recognise and appreciate the desire
to supply an accommodation which will require

and deserve the patronage and support of the
large and busy community on that line of road.

The RED ROVER will start every evening, at a
quarter before seven, by way of

Coventry,	Stafford,	Macclesfield,
Birmingham,	Newcastle-under-	and
Walsall,	Lyme,	Stockport,
	Congleton,	

and perform the journey *in the time which before
gave such general satisfaction.*

☞ It will also start from the 'Moseley
Arms' Hotel, Manchester, for London, every
evening, at nine o'clock.

> EDWARD SHERMAN, } Joint
> JOHN WETHERALD & Co, } Proprietors.

London, October 28th, 1837."

It was a gallant effort, but failed Manchester
men had grumbled at railway delays, but they
were not sentimentalists, and when the London
and Birmingham Railway was opened throughout, and an uninterrupted run through to Manchester was possible, they forsook the road, and the
" Red Rover " roved no more.

But still, sentiment gushed freely over the
coaches in every channel of the periodical press,
except, of course, in those railway journals that
even thus early had come into existence. Poetry,
of sorts, was lavished on the coachmen by the
bucketful, and they were made to consider themselves martyrs in a lost cause. They felt them-

selves greatly honoured by all these attentions,
and now began to perceive that they were really
very fine fellows indeed It was a proud position
they now occupied in the public eye, but it had
its own peculiar drawbacks. Amid all this adula-
tion they could not but see that they were like the
gladiators of ancient times, going forth to glory, it
is true, but to simultaneous extinction ; and as all
the plaudits of the multitude must have seemed
to them a hollow mockery, so did this latter hero-
worship appear cheap and unsubstantial to the
coachmen. Some of them assumed a pensive air,
which did by no means sit well upon their burly
forms and purple countenances, and was often, to
their disgust, mistaken for indigestion.

Here, from among a wealth of verse, is a
typical ballad of the time, among the best of its
kind ; but even so, perhaps not altogether one
that Tennyson would have been proud to father·—

THE DIRGE OF THE DRAGSMEN

Farewell to the Coach-box, farewell to the Vip!
By Fate most unkindly we're cotch'd on the hip,
Brother Dragsmen, come join in a general chorus,
For there's nothing at present but ruin before us.

Once who were so gay as we trumps of the team?
Now our glory hath vanish'd away, like a dream,
Doom'd to suffer adversity's punishing lash,
For the villanous Railroads have settled our hash.

Patricians no more of our craft will be backers,
And our elegant cattle must go to the knackers ;
Guards, porters, and stablemen now on a level,
And all the road innkeepers book'd for the devil

We four-in-hand worthies, however deserving,
Will have nothing in hand to prevent us from starving,
Compell'd by hard treatment our colours to strike
We may shortly turn Chartists and handle the pike.

Our beavers broad-brimm'd, and our togs out and out,
Must, the needful to raise, be soon shov'd up the spout,
Our fine, portly forms will be meagre as spectres,—
So much for these steam and these railroad projectors

By Heavens! 'tis a cruel affair, and the nation
In justice are bound to afford compensation,
And, as on the shelf we must shortly be laid,
To found an asylum for Dragsmen decay'd

There, taking our pint in all brotherly love,
We may chaff at the swells and the prads as we druv,
While spectators, admiring, exclaim'd with a shout,
" We're bless'd if that 'ere ain't a spicy turn-out '"

And how, as we tied round our necks the silk togle,
The rosy-cheek'd barmaids would tip us the ogle;
And when all was ready the ribbons to seize,
How slyly the darlings would give us a squeeze

A plague upon Railways ! the system be blowed !
Grim engineers now are the lords of the road
And passengers now are conveyed to their goal,
Not by steaming of cattle, but steaming of coal

'Tis a black, burning shame ! Must our glory be crush'd,
And the guard's lively bugle to silence be hush'd ?
Oh ! 'tis fit that our wrongs we should freely declare,
For we always look'd out for the thing that was *fare*

Let mourning as gloomy as midnight be spread
O'er the *Swan with Two Necks* and the *Saracen's Head*,
Let the *Black Bull*, in Holborn, be cow'd, and the knell
Of glory departed be heard from the *Bell*

The *Blossoms* must speedily fade from the bough,
And cross'd are the hopes of the *Golden Cross* now,
The *White Horse* must founder, the *Mountain* fall down,
The *Gloster* be clos'd, and the *Bear* be done Brown

The *Eclipse* is eclips'd, and the *Sovereign* is dead,
And the *Red Rover* now never roves from its shed;
The *Times* are disjointed, the *Blucher* at peace,
And the *Telegraph* shortly from working must cease

The *Victory* now must submit to defeat,
And the *Wellington* own he is cruelly beat,
The sport is all up with the fam'd *Tally-Ho*,
And the old *Regulator* no longer will go

Oh! had I, dear brethren, the muse of a Byron,
I'd write down the system of trav'lling on iron,
For flying like lightning but poorly atones
For crushing the carcase or breaking the bones

So, farewell to the Coach-box, farewell to the Vip!
By Fate most unkind we are cotch'd on the hip;
Then join, brother Dragsmen, in sorrowful chorus,
For at present there's nothing but ruin before us

On a few out-of-the-way routes, originally not
worth the while of railway companies to exploit,
coaching did, however, survive an incredible time.
Cordery in 1796 painted the even then old-
established Chesham coach, and coaches continued
to run into Buckinghamshire until quite recent
times. Aylesbury, Chesham, Amersham, and Wendover only obtained direct railway accommodation
when the Metropolitan Railway, under the lead of
Sir Edward Watkin, extended into the country
past Harrow and Rickmansworth, reaching Aylesbury in 1892. The Amersham and Wendover
coach—really better described as a three-horsed

'bus—went to London daily until 1890, returning from the " Old Bell," Holborn, at five o'clock in the evening. It was the sole survivor of the host of coaches that left London fifty years earlier.

But two generations have passed away since coaches began to disappear and to become historic, and the " elderly man," with his enviable memories of a long journey in mid-spring or autumn on the outside of a stage-coach, written about by George Eliot, is no longer to be found, reminiscent of the times that were. Nay, the locomotive steam engine itself is doomed, in turn, to be replaced by self-moving electric motor carriages, and we shall live to drop a salt tear upon an express locomotive retired from active service, or to sigh at sight of a solitary Metropolitan Railway engine placed in a museum of things that were. The days of the prophets were not ended with the Biblical prognosticators, with Nixon, red-faced or otherwise, or with Mother Shipton, or even with Erasmus Darwin, who, although he could foresee steam and the balloon, could not envisage electricity. They included George Eliot, also, among the prophets, shadowing forth, in a most remarkable way, the Central London Railway and other tube lines of our own time, in this extraordinary passage: " Posterity may be shot, like a bullet, through a tube, by atmospheric pressure . . . but the slow, old-fashioned way of getting from one end of our country to the other is the better thing to have in the memory. The tube journey can never lend

THE CHESHAM COACH, 1796.

From the painting by Cawley.

much to picture and narrative ; it is as barren as an exclamatory ' O ! ' " How true ! The scenery on what the vulgar call the " Tuppenny Tube " is distinctly uninteresting.

But Marian Evans had, you see, her limita- tions as a diviner of things to be. Electricity was not within her ken; she did not suspect the steam-carriages of her youth would be reincar- nated as modern motor-cars Yet, all the time, they were simply laid by, and Gurney, Hancock, and their fellows are justified in this our day Everything recurs, essentially the same as before, with a complete revolution of the wheel of time, and thus the Road has become itself again.

Will a time come when the day of the motor- car will be looked back upon with that air of regretful sentiment with which the vanished Coaching Age is regarded ? The rhythmic footfall of the horses and the rattle of the bars, the tootling of the " yard of tin " and the cheerful circumstance that attends the progress of a well- appointed coach, are things which have been, and may still be, experienced in our time by those who journey down the roads affected by the summer coaches, to Brighton, St. Albans, and Virginia Water; but as the Coaching Age itself has passed away, these are only sentimental revivals. The horseless carriages are upon us, and " going strong," alike in speed and scent. The odour of the imperfectly-combusted petrol desecrates the airs of the country-side. Already the length and breadth of the land have been explored by them, on roads good, bad and indifferent, hilly or flat; and the characteristic rattle of their machinery and the hoarse trumpeting of their cyclorns are becoming familiar even to the rustics of Devon and Somerset

Let it not be supposed, however, that skill in driving is not so necessary now as in the days of the spanking teams of coach-horses. The careful coachman of old saved his horses over the road for the long climbs and rugged places; he "sprung" them perhaps on the level, and gave them a "towelling" as a persuader to greater efforts through snow-drifts, winds or floods; and the driver of a motor-car does many of these things to his machinery, not indeed with the aid of a whip, but through the agency of levers, taps and brakes. You can overdrive and exhaust a motor just as easily as you can a horse, while it wants feeding just as well. "A just man is merciful to his beast," and a cautious man is careful of his car, not only because if he was not he would perhaps be left with half a ton of inert machinery upon the road, but because he is just as fond of his automobile as many another of his steeds of flesh and blood.

But to most people who have only seen motor-cars, and have neither driven them nor ridden in one, this will not readily be understood; while the veteran who remembers the sights and sounds of the coaching days does not hear the clatter of the new occupants of the road with pleasurable feelings. To him there is no music in the

THE LAST OF THE "MANCHESTER DEFIANCE."

From a Lithograph.

.

"Gurr-r-r-*umph*! bang, gr-rrr!" of a Daimler, changing speeds in going uphill, nor any charm in the rattle of a Benz; the "ft-ft-ft" of a motor-tricycle, or the banshee-like minor-key wail, "wow-wow-wow," of an electric cab on wood pavement. How very odd if there were!

Does it never occur to thinking men that the "blessings" of invention and the age of mechanical and other improvements have been too loudly and consistently praised? We need not be thought fanatically opposed to change if we deny the reality of some of those blessings. Let it be granted that they are ultimately in favour of the community and for the eventual improvement of the race, but if you view him unconventionally, does not the inventor, with his ingenious devices to overturn the practice and habits of generations past, seem sometimes rather a curse than a benefactor to mankind? While with one hand he simplifies and cheapens something (whether it be in travel or in anything else does not particularly matter for argument's sake), with the other he sets a more strenuous pace to life. In the long ago he invented printing, and the Devil, seeing prophetically ahead, looked on with approval, because he foresaw the halfpenny evening papers. He introduced gas, replaced horses by steam-engines, and away went the leisured pace of that generation; and then, when a newer one was born to take steam as a matter of course, brought electricity to bear upon lighting and tractive problems. Always he sets you a quicker pace when you would be going quietly or resting by the way. One generation of him takes away the traffic of the roads; another filches that of the railways and puts the traffic on the road again in an altered form. There is no finality about the inventor, who ought, for the peace of the age, first to be gently dissuaded, then admonished, and, in the last resort, severely dealt with Our ancestors had a "quick way" with such, and discouraged invention by putting inventors to death as wizards A drastic method, but they saved themselves much worry and trouble thereby. The inventor is not usually entitled to any consideration on the score of working for the benefit of humanity. So little does he do so that he takes infinite care to patent and to provisionally protect even his immature devices. He works, in short, to build his own fortune.

Apply these feelings to the case of the coachmen who were born in an age that knew nothing of steam. Every stand-by was rooted up in the coming of railways, and the steam-engine was just as strange a monster to them as the electric dynamo is to many of ourselves Often they could not transfer their allegiance to the railway, even though they starved. It was not always stubbornness or pride that held them aloof, but a certain and easily-understood lack of adaptability that forbade one who had held the reins to handle the starting-lever of the locomotive. More guards than coachmen transferred themselves from the road to the rail, because the duties were not so diverse; but, although there were coachmen who took positions on railways, no one has ever heard of one who became an engine-driver.

But coachmen and guards and the passengers they drove are all passed away, and the world rolls on as though they had never existed. The coaches, like the old Manchester "Defiance," shown in the picture, rotting away in the deserted inn-yard, were left to decay in unconsidered places or were reduced to firewood, unlike many of the old "Bull and Mouth" mails, which, after lying there for some time idle, were bought and shipped to Spain, running for many years on Peninsula roads, from Malaga in the south to Vittoria and Salamanca in the north, and by a singular fate visiting in their old age those blood-red fields of victory whose fame they had once spread from London all over triumphant England.

CHAPTER XIII

WHAT BECAME OF THE COACHMEN

"Steam, James Watt, and George Stephenson have a great deal to answer for. They will ruin the breed of horses, as they have already ruined the innkeepers and the coachmen, many of whom have already been obliged to seek relief at the poor-house, or have died in penury and want."—*The Times*, 1839

"WHERE," asked Thackeray in *Vanity Fair*, "where is the road now, and its merry incidents of life? Is there no Chelsea or Greenwich for the honest, pimple-nosed coachmen?" No, there was not. The action of Parliament in sanctioning so many railways in so short a space of time, without making any legislative restriction or provision in favour of the coachmen whose careers were ruined by railways, seems strange to the present generation, but in no single instance were they considered. The greatest and swiftest revolution ever brought about in the methods and habits of travelling took place in the short period of time between 1837, when the effect of railways first began to be felt, and 1848, when most of the great main lines were opened. Eleven years is no great space in which to effect so sweeping a change, and it is not surprising that ruin and misery were wrought by it, not among coachmen alone, but dealt out impartially to every one of the many

THE COACHMAN, 1832.   *After H. Alken.*

people and interests whose prosperity was bound up with the continuance of the old order of things. Coachmen were by no means the greatest sufferers: others felt the blow as severely, but in this chapter we have no concern with the great army of inn-keepers, ostlers, post-boys and stable-helpers who so suddenly found their occupation taken away and no new means of livelhood provided.

What became of the coachmen ? In the vast majority of cases we do not, and cannot, know; for if one thing be more certain than another, it is that we are better informed in classic and mediæval lore than in the story of our forbears of two or three generations ago, and that most of the papers and documents necessary to a full and particular history of coaching have been destroyed.

Many among those not born in the age of coaches have marvelled at what they consider the wealth of reminiscences about the old coachmen. The truth is that there exists no such wealth. There were certainly no fewer than three thousand coachmen throughout the country in the days just before railways. What do we know of them? Very little. Even their names have been for-gotten, except in some (comparatively few) special cases No one can give us a complete list of the coachmen of the Edinburgh Mail, of the Exeter "Telegraph," or Devonport "Quicksilver," or of any of the crack day coaches. Nearly complete in some cases, but never quite, because the reminiscent travellers by famous mail or stage have never troubled to detail such things; caring only to narrate the peculiarly bad or good coach-manship, as the case might be, or the eccentricities in manner or dress, of the men who drove them. The merely efficient coachman, with no salient characteristics to be described enthusiastically or spitefully caricatured, stood little chance of notice in print. He drove until the natural end of his career came, or until it was cut short by the railway; and in either case ended obscurely.

On the other hand, the noted masters of the art of driving a coach, who taught the young bloods that accomplishment, or who were excellent companions with joke and song to while the hours away, have found abundant notice; and they are the chronicles of these men that make that apparent wealth of reminiscence.

The coachmen ended, as may be supposed, very variously. A generation ago, many of the city and suburban omnibuses were driven by gloomy, purple-faced men, confirmed misanthropes, who viewed the world with jaundiced eyes, and, living in vivid recollection of the past, despised themselves, their omnibuses, and the people they drove. These were the old coachmen. The Richmond Conveyance Company, whose omnibuses in the 'sixties conveyed many Londoners between the " Goose and Gridiron," St Paul's Churchyard, and that famous riverside town, employed a number of old-time coachmen, who wore tall hats with a gold band, and were never tired of telling their box-seat passengers about the

THE DRIVER 1852. *After H. Alken.*

open-handedness of the passengers of old, and inci-
dentally that travellers by 'bus were "not worth
a d——n"; not, perhaps, a tactful or ingratiating
manner, but "out of the fulness of the heart the
mouth speaketh."

When the London and South-Western Railway
was opened to Richmond, in 1843, the first station-
master was a former coachman and coach-proprietor, and a very notable one: no less a man,
indeed, than Thomas Cooper, who had in his
time run a service of coaches between London,
Bath and Bristol, and had been landlord of that
very fine old inn, the "Castle," at Marlborough,
now and for many years past a part of Marlborough College. Cooper's varied enterprises on
the Bath Road at last led him direct into the
Bankruptcy Court. When he emerged from the
official whitewashing process, Chaplin had acquired
his line of coaches, and to that highly successful
man he became a local manager. It was Chaplin
who obtained him the position of station-master,
as doubtless he had, in his influential position
of director and chairman of the L. & S.W R.,
already found many posts on that line for
coachmen, guards, and others.

Jo Walton, the famous whip of the "Star
of Cambridge," became a messenger at Foster's
Bank in that town, after the railway had run
him off. At an earlier date Dick Vaughan, of
the Cambridge "Telegraph," had been killed by
being thrown out of a gig; but of him we know
little. Of Thomas Cross, who was intimately

connected with Cambridge, we know a good deal. He drove the Lynn "Union" for many years. Born in 1791, he died in 1877, in his eighty-sixth year. His occupancy of the box-seat lasted from 1821 to 1847, when his coaching career was brought to a close by the opening of the length of railway between Cambridge, Ely, and King's Lynn. His was a remarkable history. His father, John Cross, from being a highly prosperous coach-proprietor, with large estates and considerable social standing in the district between Petersfield and Portsmouth, was gradually brought low by misfortune and reckless speculations. John Cross, with the wealth and status of a country squire, had given his son Thomas an excellent education, and had destined him for the Navy; but serious attacks of epilepsy, and the results of an accident caused from falling in one of these fits on a number of wine-bottles, cut his career in the Service short. He was a midshipman when these distressing circumstances entirely altered his future. He then started farming, but misfortune dogged his steps As owners of horses, himself and his father fared no better, for the terrible disease of glanders broke out and quickly carried off 120 animals Eventually ruin faced the family, and Thomas Cross at last was reduced to seeking employment as a whip in the very yard once owned by his father. At the age of thirty, then, married and with a family of his own to support, we perceive him pretty thoroughly graduated in the school of

life, and already familiar with the worst blows
that adversity
could give. In
the beginning of
his coaching career
he drove the
" Union " between
London and Cam-
bridge, but at
different periods
had the middle
and the lower
ground.

He was not
altogether a genial
coachman, and
held little inter-
course with his
brethren of the
bench, to whom
he considered him-
self, as indeed he
was, superior. It
was not, however,
a judicious atti-
tude to adopt, and
those who drove
the " Star " and
"Telegraph" Cam-
bridge coaches—
Jo Walton, James
Reynolds, and others—retorted by describing him

"A VIEW OF THE TELEGRAPH":
DICK VAUGHAN OF THE CAMBRIDGE
"TELEGRAPH."

From an etching by Robert Dighton, 1809.

as an indifferent whip. Perhaps, in fact, he was, but the "Lynn Union" was never a dashing coach, and gave no opportunity of displaying the skill demanded on others.

Tommy Cross was never so pleased as when he could pick up a box-seat passenger well grounded in the classics, or interested in poetry—for poetry first, and the classics afterwards, engaged his thoughts. He drove four-in-hand all day, and when his day's work was done retired to some solitary chamber and mounted Pegasus, who carried him on the wings of the wind to the unearthly regions where dwell the spirits of Homer and Virgil. In short, he seems altogether to have lived a fine confused unpractical life, reflected to some degree in his book, *The Autobiography of a Stage-Coachman,* an interesting but formless work, so lacking in arrangement that it is difficult from its pages to gain any very clear view of his career, and actually impossible from it to discover what was the name of the Lynn coach he drove and so constantly mentions. That it was the "Union" only independent inquiries disclose. The name "Union" must in later years have taken an equivocal and prophetic meaning to poor Thomas, for, like many another coachman, he saw with apprehension railways building all over the country and running the coaches off successive roads. He knew his own turn must come, and was early seized with fears for the future. In 1843 he published, at Cambridge, in pamphlet form, some verses in imitation of Gray's *Elegy in a Country*

THE GUARD, 1832.   *After H. Alken.*

Churchyard. He called it *The Lament and Anticipation of a Stage-Coachman*. It was, indeed, a very doleful production, describing what was already happening on other roads and was presently to befall on this It is not proposed to quote the sixteen pages of this poetical effort Let two verses suffice to show at once how, if his Muse did limp unmistakably, she was not wholly destitute of descriptive force :—

> The smiling chambermaid, she too forlorn,
> The boots' gruff voice, the waiter's busy zest,
> The ostler's whistle, or the guard's loud horn,
> No more shall call them from their place of rest.

Then comes the final catastrophe ·—

> The next we heard, some new-invented plan
> Had in a Union lodged our ancient friend.
> Come here and see, for thou shalt see the man
> Doom'd by the railroad to so sad an end

The end was not yet, but the Lynn "Union" was off the road in 1847, and Cross could not obtain any form of employment on the railway. He had already, in 1846, petitioned Parliament, but without avail; and now entered upon those unhappy years in which he eked out a precarious existence on the occasional aid given him by such men as Henry Villebois, the good-hearted Norfolk sporting squire, and others who had often been passengers on the box-seat of the "Union." In those years he published several pieces in verse, generally cast in the ambitious epic form. Unfortunately, he was not the poet he thought himself, and they are rather turgid and bombastic

specimens of blank verse   He planned and wrote
a *History of Coaching*, but in the bankruptcy of
his printers the manuscript disappeared, and so
what might have proved a really valuable work
was lost. At last, in 1865, he found a home in
Huggens' College, a charitable institution at
Northfleet, founded and endowed some twenty
years earlier by a wealthy City merchant for
gentlemen reduced to poor circumstances.   This
testimony to his social superiority above other
coachmen seems to have cheered and invigorated
him amazingly, for he was a collegian at Huggens'
beneficent institution for twelve years, and lived
to be nearly eighty-six years of age.

Less fortunate was Jack Peer, or Peers, of the
Southampton " Telegraph," famous in his day, but
reduced to driving an omnibus, and thence, being
morose and quarrelsome in that position, by de-
grees to the workhouse. His unhappy situation
became known to a gentleman who had often
travelled by him in brighter times . a handsome
subscription was raised, and he was at least
enabled to end his days in quiet retirement.

A great many ex-coachmen became innkeepers
and publicans. Among these was Ambrose Pickett,
of the Brighton " Union " and " Item," who
anticipated the end of Brighton coaching in 1841,
by becoming landlord of an inn in North Street,
with the very appropriate sign of the " Coach and
Horses."

A much more famous coachman than he—Sam
Hayward, of the Shrewsbury " Wonder "—followed

Mr. Weller's example, and married a widow, landlady of the "Raven and Bell," on Wyle Cop; but
he did not long survive the extinction of "the
Road," and the widow soon found herself again in
that situation   John Jobson, who for many years
drove the "Prince of Wales"—the "Old Prince,"
as it was familiarly called—a London, Oxford and
Birmingham coach, continued on to Shrewsbury
and Holyhead—became a coach-proprietor, established at the "Talbot," Shrewsbury, and a thorn
in the side of Isaac Taylor, of the neighbouring
"Lion." Coaching came to an end at Shrewsbury
in 1842, and the name of Jobson was heard no
more.

Many coachmen were killed off the box in the
exercise of their profession, as, in the chapter on
accidents, has already been shown. A considerable number, secure in the affection of the
wealthy amateurs, many of whom they had
taught the art of driving, entered the service of
those noblemen and gentlemen, in some horsy or
stable capacity. The eighth Duke of Beaufort,
one of the Sir Watkin Williams Wynns, and
others, thus found employment for these refugees
of the road, and continually aided many more;
but something in the long overlordship they had
exercised over four horses, and a good deal more
perhaps in that hero-worship down the road, of
which Washington Irving writes, had spoiled
them. Their lives would not run sweetly in
fresh grooves. They could not, or would not, take
to new employments, and even, subsisting upon

charity, were often absurdly haughty, insolent, and insufferable Like horses, good living, coupled with little exercise, rendered them unmanageable, and they not infrequently quarrelled with the hand that fed them. "What do *you* know about throat-lashings and head-terrets?" contemptuously asked Harry Simpson, ex-coachman of the Devonport "Quicksilver," of Sir Watkin Williams Wynn, who, before him, had been holding forth to some of his guests upon the respective merits of those harnessing methods in the old coaching days. "Nothing practically," answered the good-humoured baronet; "my ideas are only ideas. But you know all about the subject: let us have the benefit of a professional view."

At this time Harry Simpson—"Little Harry," as he was called, undersized and "looking like a tomtit on a round of beef when on the driving-box"—was stud-groom to that Welsh landowner, who, from compassion, had taken him into his employ when coaching failed. "Little Harry," domineering and wilful as he was, remained in his service for thirty years, and died in 1886.

Some of the undoubted veterans of the old order lived to patriarchal ages, and when they died their obituary notices confounded many a writer who had lightly declared, years before, that the last of the coachmen was dead.

Matthew Marsh, who for many years drove the Maidstone "Times," had been a private soldier in the 11th Foot, and fought and was wounded at Waterloo. He was generally averse

THE GUARD, 1852.   After H. Alken.

from mentioning that fact, but one day, hearing
from his box a dispute about the battlefield in
which both disputants were in error, he corrected
them, simply adding, " I happened to be there."
He died in 1887, aged ninety-four years, aided
in his declining days by the Earl of Albemarle,
who had fought in the same campaign

William Clements, of Canterbury, who had
driven the " Tally-Ho " and " Eagle " coaches
between Canterbury and London before the nineteenth century had grown out of its teens, died
in 1891, aged ninety-one. He was " the last of
the coachmen," yet, two years later, in the early
part of 1893, we find the death recorded of Philip
(commonly called " Tim ") Carter, aged eighty-
eight. He it was who drove the " Red Rover "
on June 19th, 1831, from the " Elephant and
Castle " to Brighton in 4 hours 21 minutes—a
pace then greatly in excess of anything before
accomplished on that road.   The occasion was
the opening of William IV.'s first Parliament,
and the haste was for the double purpose of
speedily carrying the King's Speech to Brighton
and of advertising the " Red Rover " itself, then
a newly-established coach.   He did not run light,
as many of the record-making coaches used, but
carried fourteen passengers on that trip.

A year after Carter's death Harry Ward passed
away, August 4th, 1894, aged eighty-one.   He
was one of a family of ten, and the last, except
his elder brother Charles, of whom mention will
presently be made.   Their father had himself

been a coachman on the Exeter Road, and lived at Overton at the time Charles was born. He afterwards became landlord of the "White Hart," Hartford Bridge, on the same great highway, eighteen miles nearer London. Harry Ward's career is partly told on page 247, Vol I. In after years he drove coaches started in the revival on the Brighton Road and elsewhere.

"Last," it was again said, of the coachmen who drove the famous coaches up to the time when railways ran them off the road, was Charles S. Ward, elder brother of the above. He was born in 1810, and died in his eighty-ninth year, December 9th, 1899. His was an interesting career Son of one who had been a small proprietor as well as coachman, and thus familiar from his birth with horses, he was driving the Ipswich and Norwich Mail as far as Colchester at the early age of seventeen, and was thus probably the youngest coachman ever entrusted with the conduct of a mail on any road. But he drove it for nearly five years without an accident, and was then promoted to the Devonport "Quicksilver," at that time the fastest out of London, nightly driving the 29 miles to Bagshot, and then back, in the small hours of the morning, with the up-coach After nearly seven years of this night-work, trying and monotonous even in summer, but extremely hazardous in winter, he sought a change, and applied to Chaplin, who was the proprietor of the "Quicksilver," for day-work. The very fact

of his being so sure and safe a coachman on
the night mail operated at first against his being
transferred to a coach not calling in so great a
degree for those qualities, but in 1838 he obtained
the offer of the Brighton Day Mail, which Chaplin
was about to start, together with the chance of
horsing it a stage. Like many coachmen, ambitious of becoming a proprietor, Ward closed
with this offer, but the Day Mail did not load
well, and he soon gave up his share. He might
have known that Chaplin, so keen a business man,
was not precisely the person to offer any one else
a share worth retaining.

Ward then left Chaplin, and went over to the
Exeter "Telegraph," the fast day coach run by
Mrs. Ann Nelson, in opposition to Chaplin's
"Quicksilver Mail." Mrs. Nelson was glad to
get so steady a whip as Ward, who for three
years from this time drove the "Telegraph"
daily between Exeter and Ilminster, a double
journey of 66 miles. In 1841 the Bristol and
Exeter Railway, a continuation of the Great
Western, was opened as far as Bridgewater, and,
by consequence, the "Telegraph" was withdrawn
by Mrs. Nelson and her co-partners. Ward,
however, held on, and, with the coachman on
the other side of his stage and the two guards,
extended the journey at one end as the railway
cut it short at the other. From 1841 to
April 30th, 1844, the "Telegraph" therefore
ran the 95 miles between Bridgewater and
Devonport, taking up the railway passengers at

the former place. On May 1st, 1844, the railway
was opened to Exeter, and the journey of the
poor old "Telegraph" was cut down to 50 miles.
But those were spirited times, and even then,
driven thus into the West, there were com-
peting coaches. A "Nonpareil" Bristol and
Devonport coach had been running daily at the
same hours as the "Telegraph," but was taken
off, and a "Tally-Ho" put on the shorter Exeter
and Devonport trip. *Then* the racing became
furious. Up out of Exeter, on to the breezy
heights of Haldon, and by the skirts of Dartmoor
the two coaches sped—the "Telegraph," as Ward
tells us in his reminiscences, always leading.
Several times they did the 50 miles in 3 hours
20 minutes, and for months together never
exceeded 4 hours!

That mad pace could not last; and so, as
neither could run the other off the road, they
agreed to keep it amicably for so long as the
railway, pushing irresistibly onward, would suffer
them to exist. On May 1st, 1848, the South
Devon Railway was opened to Plymouth, and
it seemed as though coaching in the West of
England was quite killed; but a number of
Cornish gentlemen approaching Ward with the
proposal that he should start a fast coach into
Cornwall, and promising to support it, he put
a "Tally-Ho" on the road between Plymouth,
Truro and Falmouth, a distance of 62 miles.
He was so fortunate as to be offered the
contract for carrying the mail between those

places, and the "Tally-Ho" was converted into a
mail, and ran for a number of years until the
railway was opened to Truro, in May 1859.
Then, and then only, did Ward's career as a
coachman end, for although for some years, being
proprietor, he had seldom driven, he had not
hitherto deserted the box-seat, despite the calls
upon his time of the horse-mart and driving-
school business he had meanwhile established at
Plymouth.

Charles Ward, more fortunate, more businesslike and far-seeing than the majority of his
fellows, ended as the prosperous proprietor of
livery stables in the Brompton Road, in whose
yard he might be seen on sunny days during his
last years sitting on a bench against the warm
brick wall, and dozing the afternoons away.

Even as this page is written, in January 1903,
another old coachman—again "the last"!—has
died This was Sampson Brewer, who, living
in his later years at Cedar Cottage, Vancouver,
declared himself to be the last survivor of the
old coaching days. Born in 1809, he was, therefore, ninety-four years of age at his death.
He said he drove on its final journey "the last
regularly-running mail in England": that between
Plymouth and Falmouth, by way of Liskeard and
St. Austell. He must thus have been in the
employ of Charles Ward.

Two, at least, of the coachmen committed
suicide. One of these was Dick Vickers, who had
driven the Holyhead Mail. In an evil hour he

resigned the ribbons to indulge a fancy he had
nursed of becoming a farmer. But farming was
beyond him: he lost all his money at it, and
hanged himself in one of his own barns at Tynant,

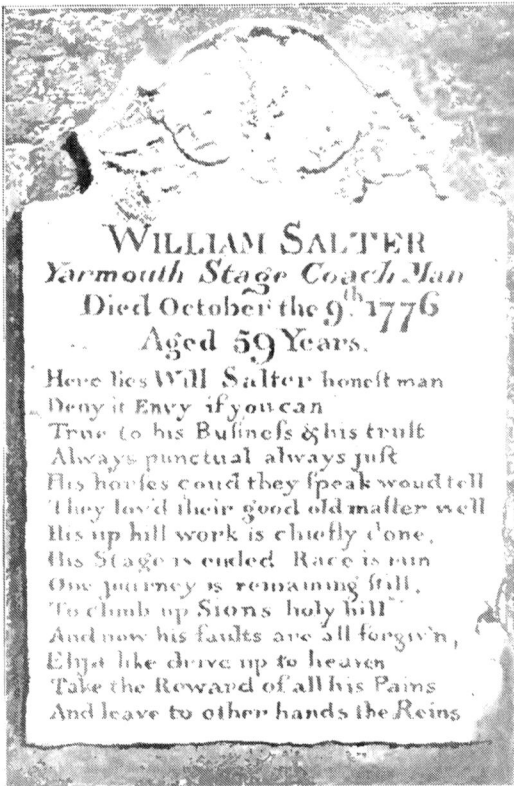

WILLIAM SALTER
Yarmouth Stage Coach Man
Died October the 9th 1776
Aged 59 Years.

Here lies Will Salter honest man
Deny it Envy if you can
True to his Business & his trust
Always punctual always just
His horses could they speak woud tell
They lov'd their good old master well
His up hill work is chiefly done,
His Stage is ended Race is run
One journey is remaining still,
To climb up Sions holy hill
And now his faults are all forgiv'n,
Elija like drive up to heaven
Take the Reward of all his Pains
And leave to other hands the Reins

A STAGE COACHMAN'S EPITAPH AT HADDISCOE.

near Corwen. Charles Holmes, for more than
twenty years coachman and part-proprietor of the
"Old Blenheim" London, Oxford and Woodstock
coach, and the recipient in 1835 of a handsome

present of silver plate, subscribed for by Sir Henry
Peyton and many other gentlemen, committed
suicide by throwing himself off a steamer into
the Thames

The question, "What became of the coachmen?" is partly answered in the subjoined col-
lection of epitaphs and eulogies got together from
far and near. First comes the early and curious
one at Haddiscoe, near Lowestoft, to William
Salter, said to have lost his life by falling from his
coach at the foot of the hill near the churchyard,
shown on the page opposite.

To this succeeds the highly interesting example
in Over Wallop churchyard, Hampshire, to Skinner,
the coachman of the Auxiliary Mail, upset at
Middle Wallop, on the Exeter Road, by one of
the wheels coming off. Skinner was killed on the
spot, and the passengers injured. The inscription
runs :—

Sacred
to the Memory of
HENRY SKINNER, a Coachman,
who was killed near this place
July 13th, 1814,
Aged 35 years

With passengers of every age
With care I drove from Stage to Stage,
Till Death's sad Hearse pass'd by unseen,
And stopt the course of my machine

Then comes a Latin passage :—

Dum socios summa per vicos arte vehebam
Mors nigra praetenit—
Machina cassa mea est.

It may be translated :—

> While I was conveying various passengers with the greatest
> skill, Black Death intervened—
> My machine is broken

An epitaph is (or was, for most of the stones in late years have been cleared away) in Winchester Cathedral yard to the last coachman of the Winchester and Southampton stage, but no record of it has been found.

Far away, in South Shropshire, on the north side of St. Lawrence's churchyard, Ludlow, lies John Abingdon, who died in 1817, and who, according to his epitaph, "for forty years drove the Ludlow coach to London; a trusty servant, a careful driver, and an honest man"

> His labour done, no more to town
>   His onward course he bends,
> His team's unshut, his whip's laid up,
>   And here his journey ends.
> Death locked his wheels and gave him rest,
>   And never more to move,
> Till Christ shall call him with the blest
>   To heavenly realms above

In the same district, in the pretty churchyard of Stanton Lacy, may be found a stone to the the memory of John Wilkes, of the Worcester and Ludlow Mail, killed in 1803 by its overturning in a flood. Some poetic friend inscribed this tribute :—

> Alas! poor Wilkes, swift down the winding hill
> The horses plunged into the fatal rill.
> The quiv'ring bridge broke down beneath the weight,
> And Wilkes was flung into the foaming spate

On his prone form the coach then t . . (? toppled) o'er,
And he was crushed beneath, to rise no more
No more to rise? No, no! Though here his work be ended,
To Heav'n we hope his spirit hath ascended.
Although on Earth his final drive be drove,
He's entered on a longer Stage above,
Where, now his mortal days are past and gone—
He drives with Phœbus' self the chariot of the Sun

Then there is the epitaph on the driver of the coach that ran between Aylesbury and London, written by the Rev. H Bullen, vicar of Dunton, in whose churchyard he is laid :—

Parker, farewell! thy journey now is ended,
Death has the whip-hand, and with dust thou't blended;
Thy way-bill is examined, and I trust
Thy last account may prove exact and just.
May He who drives the chariot of the day,
Where life is light, whose Word's the living way;
Where travellers, like yourself, of every age
And every clime, have taken their last stage—
The God of mercy and the God of love
"Show you the road" to Paradise above

The old whips had a whimsical way with them, and sometimes not a little pathetic as well. The road was not only the profession whence they drew their living, but it was their passion—their whole life. Thus, when a noted chaise-driver at Lichfield, one Jack Lewton, died in 1796, he was, at his last request, carried from the "Bald Buck" in that city by six chaise-drivers in scarlet jackets and buckskin breeches—the pall supported by six ostlers from the different inns. The funeral took place on August 22nd, in St. Michael's churchyard, as near the turnpike road as possible; so that he might, as he said, enjoy the satisfaction of hearing his brother whips pass and repass.

Similar directions are said to have been left by Luke Kent, reputed to have been the first guard ever appointed to a mail-coach. The story goes that he was buried at Farlington, near Portsmouth, on the Chichester Road, and left an annual bequest to his successors on the Chichester coach, on condition that they should always sound their horns when passing the place of his interment. Diligent inquiry, however, does not disclose the fact of any one of that name lying at Farlington; but a Francis Faulkner, who died at Petersfield, May 18th, 1870, aged eighty-four years, lies in a vault in Farlington churchyard. He was a guard on the "Rocket" London and Portsmouth coach, and local gossip still tells that he left a request (perhaps also a bequest) that if ever stage-coaches should pass his vault, their horns should be sounded. Certainly, a few years ago, when a coach was run from Brighton to Portsmouth, its horn was always sounded on passing the churchyard.

A conclusion shall be made with the eulogy of Robert Pointer, coachman on the Lewes stage, which he is said to have driven thirty years without an accident. It does not appear what relation he was to the one-time famous "Bob Pointer," of the Oxford Road, and in 1834 on the Brighton "Quicksilver"—a favourite coaching tutor. *That* Bob Pointer, according to the Duke of Beaufort, could always be depended on to

start sober, but the horses had to be changed on
the way anywhere but at public-houses, if it was
desired that he should end his journey in the
same condition :—

> Those who excel, whatever line 'tis in,
> Deserve applause, and ought applause to win
> Pointer in coachmanship superior shone,
> His whip his sceptre, and his box his throne.
> Not skilled alone the fiery steeds to guide,
> For them in sickness and in health provide,
> He, by a thousand nice *minutiæ*, knew
> To win the restive, and the fierce subdue,
> As man and master, punctual and approved
> By those who knew him best, the best beloved.
> Many's the time and oft, o'er Ashdown's plain,
> 'Mid show'rs of driving snow and pelting rain;
> When hurricanes bow'd down the lofty grove,
> When all was slough beneath and storms above;
> And oft, when glowing skies cheer'd all the scene
> And threw o'er Sussex plains a joy serene,
> When now the anecdote, and now the song
> Beguil'd the moments as we roll'd along:
> Snug at his elbow have I mark'd his skill
> To rein the courser and to guide the wheel;
> And had he Phaeton's proud task begun,
> To drive the rapid chariot of the sun,
> Safe through its course the flaming car had run.

CHAPTER XIV

THE OLD ENGLAND OF COACHING DAYS

THIS is the time, now that we have passed the
threshold of a new era, when old landmarks are
disappearing everywhere around us as we gaze,
and the Old England that we have known is
being dispossessed and disestablished by a new
and strange, an inhospitable and alien England
of foreign plutocrats—this is the psychological
moment for a brief review of what this England
of ours was like in the old days of stage-coach
and mail

If we could recapture those times we should
find them spacious days, of much fresh air,
illimitable horizons, a great deal of solid, un-
ostentatious comfort for the stay-at-homes, and
also of much discomfort for the traveller, but
although no sensible person, fully informed of
the conditions of life in the long ago, would wish
he had been born into those times, yet among
their disadvantages and the discomforts incidental
to travel scarce more than two generations ago,
there were to be found, as a matter of course, not
a few things which would be looked upon with
rapture by the modern sentimentalist. That was
the era when the Suburb was unknown anywhere

else than around London, and even London's
suburbs were sparse, scattered, sporadic, and
separated by great distances from one another.
Taking coach from the City, where the merchants
and the shopkeepers commonly lived over their
business premises, you came presently, north,
south, east, or west, through suburban Stamford
Hill, Sydenham, Clapton, or Kensington, to rural
Edmonton, Croydon, Romford, or Chiswick, and
so presently to the Unknown. *That* was, of
itself, a charm in the old order of things—a
charm lost long since in these crowded times,
when constant and intimate travel have made
us familiar with distant towns, and by consequence incurious and incapable of surprises.
Everything is known, if not at the first hand
of personal observation, at least by proxy of
our reading in guide-book history, or by the
debilitating photograph, which leaves nothing to
the imagination, and renders us travelled in the
uttermost nooks and corners of the land, even
though we be bedridden, or thoroughgoing
habitués of the armchair and the fireside. The
picture-postcard—the lowest common denomi-
nator of the photograph—has come to give the
last touch of satiety, the final revulsion of re-
pletion. The Land's End has long since been
exploited, John o' Groat's is merely at the end
of a cycle ride, the "bottomless" caverns of
the Peak have been plumbed, every unscalable
mountain climbed. "*Connu !*" we exclaim when
we are told any fact No surprises are left.

We may never before have journeyed to Edinburgh, but photographs have rendered us so long familiar with its castle and rock that we cannot recollect a time when we were not familiar with the physical geography of the "modern Athens," and we seem to have been born with a knowledge of the geographical peculiarities of every other place. We are, therefore, naturally bored and unresponsive in situations where our grandfathers were surprised and delighted; but although possessed thereby with a profound dissatisfaction with ourselves, we cannot hope to win back to the unsophisticated joys of old time.

Would that it could be done! The wish is everywhere evident, but only Lethean waters could sweep away the useless lumber of mental baggage that destroys imagination and blunts the senses. The many efforts made to bring back the "properties"—to speak in the theatrical sense—of old time are pitiful or ridiculous, as your humour wills it. These are the days when things quaint and old-fashioned are revived for sake of their quaintness, sometimes in spite of their inconvenience and unsuitability; when ingle-nooks and open hearths with fire-dogs are built into modern houses for effect, although slow-combustion stoves are infinitely more comfortable and less wasteful of fuel. Our forbears, who did not know slow-combustion stoves, were not the creatures of sentiment that we are, and would soon have abolished open hearths for the

close stoves had they been given the chance,
just as they would have exchanged the tallow
dip for electric lighting had the opportunity
offered. We do not know the feelings with
which the first gentlemen to use carpets abolished
the old rush-strewn halls and the manners and
customs contemporary with them; but if their
sense of smell was as acute as our own, they must
have noticed with great relief the absence of the
dirt and festering bones that found a hiding-place
beneath those rushes All the marvellous changes
in habits of living—the cheapening of food,
the conversion of the luxuries of a former age
into the ordinary requirements of this, and even
the alterations in the face of the country and the
houses of towns and villages—are due to those
increased facilities of intercourse which, owing to
the gradual improvement in roads, the coaches
and waggons of yore were first able to give.
When public vehicles began to ply into the
country, this England of ours was not only a land
of wide unenclosed heaths and commons, but the
people of one county—nay, even the inhabitants
of towns and villages—were markedly different
in thought and prejudices, in speech and clothing,
from those of others; while local style in building,
and the various building materials obtained locally,
gave each successive place that appearance of
something new and strange which the traveller
does not always meet with nowadays in far distant
lands. As the drainage of lakes and fens, the filling
up of the valleys and the reduction of the hills,

have quite revolutionised the physical geography
of wide areas, often changing the natural history
of the districts affected, so has cheap, constant
and quick travelling and conveyance of materials
helped to reduce places and people to one dead
level. Romance flies abashed from the level,
monotonous road, where, years before, in some
darkling hollow between the hills, ringed in by
dense woodlands, it lurked in company with the
highwayman. We do not desire the return of
those gentry, but what would literature have done
without them? Highway and turnpike improvements long ago sliced off the most aspiring hilltops, and, carrying the roads through cuttings,
used the material thus cut away for the purpose
of filling up the gullies and deep depressions.
Where the early coaches toiled, often axle-deep,
through the watersplashes formed by the little
rills and streams that ran athwart the way, later
generations have built bridges, or have done things
infinitely worse; so that a watersplash has become
a rare and curious object, noteworthy in a day's
journey. Only recently, on the Dover Road, near
Faversham, has such a watersplash—one of the
most picturesque in the country—been abolished.
Ospringe was a little Kentish Venice, with a
clear-running shallow stream occupying the whole
of the roadway, with raised footpaths for pedes-
trians at either side, and ancient gabled cottages
looking down upon the pretty scene. Alas! the
sparkling stream now goes under the road, in a pipe.

In the old days, no traveller going north along

the Great North Road left Alconbury without first seeing that the priming of his pistols was in order, while the passengers by mail or stage secretly put their watches and jewellery between their skin and their underclothing, or deposited their purses in their boots, before the coach topped Alconbury Hill. For at "Aukenbury," as Ogilby in his old road-maps styles it, you were on the threshold of a robbing-place only less famous than Gad's Hill, near Rochester, or those other notorious dark or daylight lurks (for day or night mattered little in those times), Hounslow Heath and Finchley Common. The name of this ill-reputed place was "Stonegate Hole." It is marked distinctly on the maps of Ogilby and his successors, between the sixty-fourth and sixty-fifth milestones from London, by the Old North Road, measured from Shoreditch, and passing through Ware, Royston, and Caxton   .

Passing Papworth Everard, you came in those days, on the left hand, just before reaching the fifty-sixth milestone, to " Beggar's Bush," where you probably saw the tramps, vagrants and foot-pads of that age skulking, on the chance of robbing some traveller unable to take care of himself. Here, in sight of these wretches, you ostentatiously toyed with your pistol holsters, or loosened your sword in its scabbard, and so passed on scathless. On leaving Alconbury, however, the horseman generally preferred company, because the highwaymen of Stonegate Hole were well armed, and, by consequence, courageous.

What, exactly, was Stonegate, or Stangate, Hole? It was the deep and solitary hollow that then existed at the foot of the northward slope of Alconbury Hill, known now as Stangate Hill. The name derived from this road being a part of the old Roman "Ermine Street," formerly a stone-paved way, and the "Hole" was formed by a rise that immediately succeeded the descent. Quite shut in by dense woods, it was an ideal spot for highway robbery. When, in the later coaching era, the road was lowered through the crest of the hill, and the earth was used to raise it in the hollow, Stonegate Hole disappeared. Bones were found during the progress of the works, supposed relics of unfortunate travellers who had met their death at the hands of the highwaymen. A more or less true story was long told of an ostler ot the "Wheatsheaf," the inn that once stood on the hill-top. He, it seems, used to help in putting in the coach-horses when the teams were changed, and would then take a short cut across the fields, and be ready for the coach when it came down the road. The coachman, guard, and passengers, who did not know that the shining pistol-barrel he levelled at them was really a tin candlestick, were duly impressed by it, and yielded their valuables accordingly.

A tale used to be told of one of the old "London riders," or "bagmen," who lay at the "Wheatsheaf" overnight and set forth the next morning. His saddle-bags were full, and so weighted with samples of his wares that he could scarce sit his horse, and had to be helped into the saddle by an ostler. Once up, his eyes only with difficulty peered over this mountainous weight, but in this manner he set forth. He had not gone far before he thought he had lost his way, when fortunately he perceived another horseman, and hailed him. The stranger took no notice; and so our traveller ranged up alongside him with the question. Instead of replying, the stranger thrust his hand into his breast-pocket and withdrew what the traveller imagined to be a pistol. Recollections of the evil repute of the place suddenly rushed into the traveller's mind, and, putting spurs to his horse, he dashed away from the supposed highwayman, and did not draw rein until in the neighbourhood of Huntingdon.

There he met a party of horsemen, who determined to hunt the highwayman down, and so, with the traveller, hurried on to Stonegate "There he is!" cried the traveller, as they came in view of a peaceful-looking equestrian, ambling gently along.

"You are mistaken, sir," said one of the party: "that is our Mayor, the Mayor of Huntingdon"

But the bagman asserted he was right, and so, to end the dispute, the whole party rode up, and one wished "Mr. Mayor" good morning. It was indeed that worthy man, and although he again, instead of making answer, drew something from his pocket, it produced no alarm among his fellow-burgesses, for *they* at least knew him for a very deaf man, and had often seen him reach for that ear-trumpet which he now drew forth, clapped to his ear, and asked them what it was they said.

Swift, who, travelling between London, Chester, Holyhead and Dublin, remarked upon the many nations and strange peoples he passed on the way, serves to emphasise these notes upon the fading individuality of places and people. The dialect of "Zummerzet" has not wholly decayed, but it has become so modified that when old references to its Bœotian nature are found, the reader who knows modern Somerset, and does not consider these changes, concludes that its grotesque speech was greatly exaggerated; just as he cannot be made to implicitly believe the remarkable and oft-repeated story told by William Hutton of the visit of himself and a friend to Bosworth in 1770, when the people set the dogs at them, for the only reason that they were strangers, or that other tale of the savagery of the Lancashire and York-shire villagers, who, when a person unknown to them appeared, conversed as follows —

"Dost knaw 'im ? "

"Naya. '

" Is't a straunger ? "

" Ay, for sewer."

"Then pause 'im, 'eave a stone at 'un; fettle 'im "

No inoffensive stranger in country districts is likely to meet with that reception nowadays. The stranger in those times was regarded, as he gene-rally is in savage countries, as necessarily an enemy, but travel has changed all that, and it has been reserved for the London "hooligan," who has been taught better, to perpetrate, in the very centre of civilisation, the barbarous methods of the uninstructed peasantry of generations ago

Stories like these are only incredible when the circumstances of the age are unknown. In times when a stranger might easily enough prove to be a highwayman, or at the very least some Government emissary intent upon collecting hearth-money, window-tax, or one of the very many duties then levied upon necessaries of life, a strange face might be that of an enemy, and at any rate was unlikely to be that of a friend. Sightseers were unknown. No one stirred from home if he could find an excuse for staying by his own fireside. "What do you want here?" asked the Welsh peasants of the earliest tourists; and declined to believe them when they said they journeyed to view the Welsh mountains. "For Christianity's sake, help a poor man!" implored an early traveller in Scotland, fainting by the way. The door was slammed in his face. "Surely you are Christians?" exclaimed the unhappy man. "There are no Christians here," replied the half-savage Scot: "we are all Grants and Frasers." That last is, perhaps, rather a savagely humorous than a true story, but the mere existence of it is significant. More authentic —nay, well established—is the statement that even so late as 1719, in Glasgow, two people of the same name would commonly be distinguished by some physical peculiarity; or else, if one was travelled and the other not, the one who had been to the capital would be " London John," or James, according to what his Christian name might be.

A course of reading in the " travels " of the authors and diarists who ambled about England, on horseback or otherwise, in the old days, sufficiently demonstrates the aloofness and isolation, and the essential differences that divided the country districts. When the Dukes of Somerset resided at Petworth, in Sussex, the roads were so bad that it was next to impossible to get there, and when once there it was equally difficult to get away. Petworth is only forty-nine miles from London, but the Duke of Somerset maintained a house at Godalming, sixteen miles along the road, where he could halt on the way and pass the night. His steward generally advised the servants some time before his Grace started, so that they might be on the road " to point out the holes." When the Emperor Charles VI. visited Petworth, his carriage was attended by a strong escort of Sussex peasants, to save it from falling over. In spite of their efforts, it was several times overturned, and that was a very sore and bruised Emperor who supped that night with the Duke Similar adventures befel Prince George of Denmark, husband of Queen Anne, visiting Petworth from Windsor He went in some state, with a number of carriages. " The length of way was only forty miles, but fourteen hours were consumed in traversing it, while almost every mile was signalised by the overturn of a carriage, or its temporary swamping in the mire. Even the royal chariot would have fared no better than the rest, had it not been for the relays of peasants who poised and kept it erect by strength of arm, and shouldered it forward the last nine miles, in which tedious operation six good hours were consumed."

The travellers of that era, knowing how strange the country must be to most people, gravely and at length described places that in these intimate times an author would feel himself constrained to apologise for mentioning, except in a personal and impressionistic way, and they not only so describe them, but there is every reason to believe their writings were read with interest. More interesting than their dry bones of topographical history are the accounts they give of manners, customs, and thoughts common to the time when travellers were few and little understood. When, in 1700. the Reverend Mr. Brome, rector of the pleasant Kentish village of Cheriton, determined to make the explorations of England that took him, in all, three years, he was obliged, as a matter of course, to wait until the spring was well advanced and the roads had again become passable. Setting forth at last, one mild May day, his friends and parishioners accompanied him a few miles, and then, with the fervent "God be with you's" that were the parting salutations of the time, instead of the lukewarm "Good-bye's" of to-day, turned back home-along, and expected to hear of him no more. But he *did* return, as his very dull and jejune

book, chiefly of stodgy historical and topographical information, published in 1726, sufficiently
informs us.

"Weeping Cross" is the name of a spot just
outside Salisbury, supposed to have taken its
name from being the spot where friends and rela-
tives took leave of travellers, with little prospect
in their minds of seeing them again. There is
another "Weeping Cross" on the London side of
Shrewsbury, near Emstrey Bank, about a mile
from the town and overlooking the descending
road, whence the progress of the travellers could
be followed until distance at last hid them from
view. There are, doubtless, other places so
named throughout the country. The oft-repeated
legendary statement that travellers usually made
their wills before setting out is thus seen to be
reasonable enough, but it is specifically supported
by the author of *Letters from a Gentleman in the
North of Scotland*, who, writing about 1730, says :
" The Highlands are but little known, even to the
inhabitants of the low country of Scotland, for
they have ever dreaded the difficulties and dangers
of travelling among the mountains, and when
some extraordinary occasion has obliged any one
of them to such a progress, he has, generally
speaking, made his testament before he set out,
as though he were entering upon a long and
dangerous sea-voyage, wherein it was very doubtful if he should ever return."

When Mrs. Calderwood, of Polton and Coltness,
made a journey from Scotland into England in

1756, she wrote a diary, a very much more entertaining and instructive affair than the Reverend
Mr. Brome's book—which, indeed, could have been
compiled from other works without the necessity
of travelling, and, but for a few fleeting glimpses
of original observation, actually gives that impression. Mrs. Calderwood tells us that at Durham
she went to see the Cathedral, where the woman
who conducted her round the building did not
understand her Scottish ways (nor indeed did Mrs.
Calderwood comprehend everything English). " I
suppose, by my questions, the woman took me for
a heathen, as I found she did not know of any
other mode of worship but her own; so, that she
might not think the Bishop's chair defiled by my
sitting down in it, I told her I was a Christian,
though the way of worship in my country differed
from hers." Mrs Calderwood, quite obviously,
had never heard of St Cuthbert and his antipathy
to women, so respected at Durham that woman-
kind were not admitted within certain boundaries
in his Cathedral church ; nor was she familiar
with hassocks, for she narrates how the woman
" stared when I asked what the things were that
they kneeled upon, as they appeared to me to be
so many Cheshire cheeses."

The modern tourist along our roads finds a
deadly sameness overspreading all parts of the
country. The same cheap little suburban houses
of stereotyped fashion, built to let at from £25 to
£30 a year, that sprawl in mile upon mile on the
outer ring of London, are to be found—nay, are

insistently to the foreground—wherever he goes. They form the approach to, the outpost of, every town, large or small, he enters, and are built in the same way, and of the same materials, whether he travels farther north, south, east, or west. It was not so, need it be said, in the old times. Then the coach passenger with an eye for the beautiful and the unusual had that sense abundantly gratified along almost every mile of his course, for when men did not build on contract, and when the contractor, had he existed, would not have been able to work outside his own district, there was individuality in building design. We all know the truth of the adage that "variety is charming," and of variety the travellers had their fill. And not only was there variety in design, but an endless change of materials gratified the eyes of those who cared for these things. London, with its dingy brick, was succeeded, as one penetrated westwards, by the weather-boarded cottages of Brentford and Hounslow, by the timber framing and brick nogging of the next districts, by the chalk and flint of Hampshire and Wilts; and at last, when one had come to the stone country, by the yellow ferruginous sandstone of Ham Hill, that character-ises the houses and cottages between Shaftesbury, Crewkerne and Chard. Coming into Devon, the yellow stone was replaced by the rich red sandstone, or the equally red "cob" of that western land; and a final change was found when, the Tamar passed and Plymouth left behind, the massive granite churches, houses and cottages astonished the new-comer to those parts. No one could build with other than local materials in those days. The material might be, like the granite, stubborn and difficult, and expensive to work, but it would have been still more expensive to bring other materials to the spot, and so the local men worked on their local stone, and in course of time acquired that peculiar mastery of it and that way of expressing themselves which originated that "local style" whose secret is so ardently sought by modern architectural students. You cannot transplant the old style of a locality. Like the wilding plucked from its native hedgerow, it dies, or is cultivated into something other than its original old sweet self and becomes artificial. Cynic circumstance has so decreed it that, while these ancient local growths have in modern times been copied in London and the great towns, the rural neighbourhoods have been cursed with an ambition to copy London, while everywhere cheap red brick is ousting the native stone, flint, or wood.

When the fashionables travelled down by coach to Bath, one might safely have offered a prize for every brick house to be found there, for Bath was, and is, built of the local oolite known as "Bath stone." The prize would never have been claimed; but something like a modern miracle is now happening, for even at Bath red brick has underbid the native stone and gained an entrance.

Nothing escapes the modern desecrating touch. "Auld Reekie" itself — Edinburgh, that last stronghold of the Has Been—is not the same "beloved town" that Sir Walter Scott knew. The French Renaissance character of its grandiose new buildings does not alone tend to change it into something alien to sentiment and ancient recollection ; but that which our ancestors would have thought a mere impossibility, that which themselves would, and ourselves should, stigmatise as a crime committed against History and the Picturesque, has almost come to pass. In short, the deep ravine where the Nor' Loch stagnated of old, where the Waverley Station is now placed, has been deprived of something of its apparent depth, and the Castle Rock of a corresponding height, by the towering proportions of the vast buildings that fill up the valley and desecrate the site of the northern capital.

Sturdy survivals of olden days are the local delicacies that first obtained a wider fame from that time when they were set before the coach passengers at the country inns where the coach dined, or had tea, or supped, and were so greatly appreciated that supplies were carried away for the benefit of distant friends. Some, however, of these delicacies have disappeared. No longer does Grantham produce the cakes mentioned by Thoresby in 1683 Grantham, he says, was "famous in his esteem for Bishop Fox's benefactions, but it is chiefly noted of travellers for a peculiar sort of thin cake, called ' Grantham

Whetstones.'" What precisely were the cakes known by this unpromising name we cannot say, for the making of them is a thing of the past.

Stilton cheese, never made at Stilton, obtained its name exactly in the manner already described It was a cheese made at Wymondham, in Leicestershire, but its merits were first discovered by the coach-parties who dined at the "Bell" at Stilton, whose landlord obtained his supply from Wymondham, and drove a roaring trade in old cheeses sold to the coaches to take away. "Stilton ' cheese is now only a conventional name, like that of "Axminster" carpets, made nowadays at Kidderminster.

To bring home with him bags and boxes of local delicacies was to the old coach-traveller as much an earnest of his travels as the bringing back of a storied alpenstock is to the tourist in Switzerland. The Londoner, returning home from Edinburgh, could come back laden with a number of things which, easily obtainable now, were then the spoils only of travel. From Scotch shortbread the list would range to Doncaster butterscotch, York hams, Grantham gingerbread, and Stilton cheeses. On other roads he might secure the cloying Banbury cake, still extant, and as sickly-sweet and lavish of currants as of yore, the famous Shrewsbury cakes, manufactured by the immortal Pailin, who left his recipe behind him, so that the cakes of Shrewsbury still continue in the land; Bath buns, phenomenally adhesive and sprinkled with those fragments of loaf sugar without which the exterior of no Bath bun is complete; the cheese of Cheddar; the toffee of Everton; pork pies from Melton Mowbray; or a barrel of real natives from Whitstable. All or any of these, I say, he might carry home with him, while few places were so unimportant in this particular way that he could not ring the changes on gastronomic rarities as he went.

All these things were the products of that old English tradition of good cheer and hospitality which lasted even some little way into the railway age. Journeys were cold, but hearts were warm, and the more rigorous your travelling the better your welcome. It would seem, and actually be, absurd to surround a modern arrival by railway with the circumstance that greeted the advent of the coach. In the bygone times the guest had no sooner alighted at his inn and proceeded to his room than a knock came at his door, and lo! on a tray a glass of the choicest port or cordial the house contained. To this day the courteous old custom survives at the "Three Tuns," in Durham, whose traditional glass of cherry brandy is famous the whole length of the great road to the north.

For the little folks who travelled by coach, either with their own people or, like Tom Brown, in charge of the guard, warm motherly hearts beat in the bosoms of the stately landladies of the age, all courteous punctilio to their grown-up guests, but sympathy itself to the wearied youngsters. Such was Mrs. Botham, of the "Pelican," at

"ALL RIGHT!" THE BATH MAIL TAKING UP THE MAIL-BAGS.

From the contemporary lithograph.

Speenhamland, on the Bath Road—that "Pelican" of whose "enormous bill" some waggish poet had sung at an early period. Mrs. Botham, an awesome figure—like Mrs. Ann Nelson, of the "Bull," Whitechapel, dressed in black satin—unbent to the youngsters, for whom, indeed, she had always ready a packet of brandy-snaps.

The earlier travellers were even more welcomed, not by the innkeepers alone, whose welcome was not altogether altruistic, but by the country folk in general.

The annual reappearance of the early stage-coaches was a much greater event to the villagers and townsfolk of the more remote shires than we moderns might suppose, or feel inclined to believe, without inquiry But we must consider the winter isolation of such places in those remote times, and then some faint glimmering sense of their aloofness from the world will give us an understanding of the relief with which they again saw real strangers from the outer world. In the long winter months, when days were short and roads only to be travelled by the most daring horsemen, spurred to the rash deed only by the most urgent necessity, the passing stranger was rare, and excited remark, and the company in the inn parlour or by the ingle-nook discussed him, both because of his rarity and by reason of their own raw material for the making of conversation being run very low indeed. We should be more thankful than we generally are that our lot was not cast in a seventeenth-century village, for winter in such surroundings was dulness incarnate. Because they could not obtain fodder to keep the sheep and cattle in good condition through the winter, the farmers and graziers of that time killed them before that season set in, and the villagers lived upon salted meat   Every house had its salt-beef tub and its bacon-cratch under the kitchen ceiling, well stocked with hams and sides ; but vegetables were so scarce as to be practically unobtainable.

Every household brewed its own beer and kept a stock of cider, and most housewives were cunning in the preparation of metheglin, a sickly-sweet and heavy drink that revolts the modern palate, but was then greatly appreciated   Evenings were not long, even though it grew dark before four o'clock, for folks went to bed by seven or eight. There was little inducement to sit up late, because only the feeblest illumination was possible to any but the very rich, and the yeomen, the farmers and the cottagers had to rest content with the dim sputtering glimmer of the tallow dips that every eight or ten minutes required the attentions of the snuffers. "When the night cometh," we read in the Bible, "no man can work"; but that is a statement which, literally true at the time when the Bible was done into English, can now only be read and understood figuratively.   No one could work by the artificial illumination then possible

Conceive, then, the joy with which returning spring was greeted—spring, that brought back light and fresh food and intercourse with the

world outside the rural parish. Mankind had
travelled far from those prehistoric times of
annual terror, when the ignorant savage saw the
sun's light going out with the coming of winter,
and so, with abject fear, passed the darkling
months until the vernal solstice brought him
hope again. No one in the Old England of two
hundred and fifty years ago trembled lest the sun
should not return at his appointed time; but when
the sap rose and the birds began to sing again,
and warmth and light had begun to replace the
fogs and mists of winter, the hearts of all rejoiced.

May Day was then the great merrymaking
festival, but the first coach that ventured along
the roads, now beginning to set after the winter's
rains, had a welcome of its own. At Sutton-on-
Trent, on the Great North Road, the springtide
custom of welcoming the early coaches was
royally observed, and kept up for many years.
No coach, during a whole week of jollity, was
suffered to proceed through that jovial village
without it halted and ate and drank as only
Englishmen could then drink and eat. Guards,
coachmen and passengers were freely feasted,
willy-nilly Young and old plied them with the
good things, spread out upon a tray covered with
a beautiful damask napkin, and heaped with
plum-cakes, tartlets, gingerbread, and exquisite
home-made bread and biscuits; while ale, currant
and gooseberry wines, cherry brandy, and occasionally spirits, were eagerly pressed upon the
strangers. Half a dozen damsels, all enchanting

young people, neatly clad, rather shy, but courteously importunate, plied the passengers.

Thoresby records a similar custom at Grantham, near by, on one of his journeys. Under date of May 4th, 1714, he says · " We dined at Grantham, and had the usual solemnity, being the first passage of the coach this season , the coachman and horses decked with ribbons and flowers, and the town music and young people in couples before us." The " town music " was what we should nowadays call the Town Band.

When such courtesies obtained along the roads the coachmen and guards would have been churlish not to have, in some prominently visible manner, done honour to the season. And, indeed, May Day and springtime decorations were features on most coaches. The coachman's whipstock was ornamented with gay ribbons and bunches of flowers, while the coachman himself wore a floral nosegay that rivalled a prize cabbage in size. The guard was no less remarkable a figure, and his horn was wreathed with the most lively display of blossoms. Festoons of flowers and sprays of evergreens so draped and covered the coach that the insides, peering out upon the festivities, very closely resembled those antic figures, the " Jacks-in-the-Green," that used on May Day to prance and make merry from the midst of an embowering canopy of foliage, even so late as thirty years ago, in London streets. The horses, too, bore their part Their new harness and saddle-cloths, the rosettes and wreaths of laurel on their heads, smartened them up so that even the animals themselves were conscious of the occasion, and bore themselves with becoming pride.

Those old customs are, as a matter of course, gone. Coaches no longer dash through the old "thoroughfare" villages; and when, with the advent of spring, the motorist appears upon the road, the villagers, rather than welcoming his appearance, curse him for the clouds of dust he leaves behind. Motor-cars, they tell us, are to repeople the old coaching-roads, whose prosperity is, through them, to return, and the picturesque old wayside inns, with their memories of the coaching age, are to once again experience the rush of business. It may be so, but no one will regret the fact more than the lover of Old England, who, in the repeopling of the roads, sees their modernising inevitable, and the equally inevitable bringing "up to date" of those quaint, quiet, and comfortable hostelries so dear to the genuine tourist. It is true, they do not dine you elaborately—as your extravagant motorist complains—but life is not all chicken and champagne, and it will be a sorry day when the plain man, fleeing the gaudy glories of hotels at fashionable resorts, finds the unsophisticated inns of the countryside remodelled on the same plan. Already the picturesqueness of the old roads is threatened. They are, if you please, too hilly, too narrow, or not straight enough for that new tyrant of the highways, the owner of a high-powered motor-car, and plans have actually been drawn up by irresponsible busybodies for straight and broad new tracks, or for the remodelling of the old roads on the same principle. Roadside trees and avenues keep the surface damp and muddy after rain, and so, as rubber-tyred cars are apt to skid and side-slip on mud, the same voices call for the abolition of wayside trees. Old England is in a parlous state, when these things can be advocated and no indignant protests rise.

CHRONOLOGICAL SUMMARY

1610   Patent granted for an Edinburgh and Leith waggon-coach.

1648   Southampton weekly stage casually mentioned.

1657.   Stage-coaches introduced the London and Chester Stage.

1658.   First Exeter Stage.

  ,,   ,, York and Edinburgh Stage

1661.   ,, Oxford Stage

  ,,   Glass windows first used in carriages the Duke of York's carriage

1662.   Only six stage-coaches said to have been existing.

1665.   Norwich Stage first mentioned.

1667   Bath Flying Machine established.

  ,,   London and Oxford Coach in 2 days, established.

1669.   ,, ,, ,, Flying Coach, in 1 day, established.

1673.   Stages to York, Chester, and Exeter advertised.

1679   London and Birmingham Stage, by Banbury, mentioned

1680.   "Glass-coaches" mentioned

1681   Stage-coaches become general: 119 in existence.

1706.   London to York in 4 days.

1710 (about)   Stage-coaches provided with glazed windows.

1730.   "Baskets" or "rumble-tumbles" introduced about this period

1734   Teams of horses changed every day, instead of coaches going to end of journey with same animals.

  ,,   Quick service advertised Edinburgh to London in 9 days.

1739. According to Pennant, gentlemen who were active
    horsemen still rode, instead of going by coach.
1742. London to Oxford in 2 days
  ..   „   „ Birmingham, by Oxford, in 3 days
1751   „   „ Dover in 1½ days
1753   Outsides carried on Shrewsbury Stage.
1754   London and Manchester Flying Coach in 4½ days.
  .,   Springs to coaches first mentioned. the Edinburgh
    Stage.
  „   London and Edinburgh in 10 days.
1758. London and Liverpool Flying Machine in 3 days
1760   „   ., Leeds Flying Coach advertised in 3 days:
    took 4
1763. London and Edinburgh only once a month, and in
    14 days.
1776. First duty on stage-coaches imposed
1780   Stage-coaches become faster than postboys.
1782   Pennant describes contemporary travelling by light
    post-coaches as " rapid journeys in easy chaises,
    fit for the conveyance of the soft inhabitants of
    Sybaris "
1784. Mail-coach system established.
1800 (about). Fore and hind boots, framed to body of
    coach, become general.
  .   Coaches in general carry outside passengers.
1805. Springs under driving-box introduced.
1819. "Patent Safety ' coaches come into frequent use, to
    reassure travelling public, alarmed by frequent
    accidents.
1824. Rise of the fast day-coaches: the Golden Age of
    coaching.
  .,   Stockton and Darlington Railway opened   first
    beginnings of the railway era.
1830. Liverpool and Manchester Railway opened: coaching
    first seriously threatened.
1838. London and Birmingham Railway opened · first
    great blow to coaching, coaches taken off Holy-
    head Road as far as Birmingham

1839. Eastern Counties Railway opened to Chelmsford
1840 Great Western Railway opened to Reading.
  ,, London and Southampton Railway opened to Ports-
      mouth coaches taken off Portsmouth Road.
1841. Great Western Railway opened to Bath and Bristol :
      coaches taken off Bath Road.
  ,, Brighton Railway opened : coaching ends on Brighton
      Road.
1842. Last London and York Mail-coach.
1844 Great Western Railway opened to Exeter last
      coaches taken off Exeter Road.
1845. Railways reach Norwich
  ,, Eastern Counties Railway opened to Cambridge
1846. Edinburgh and Berwick Railway opened
1847. East Anglian Railway opened to King's Lynn.
1848. "Bedford Times," one of the last long-distance
      coaches withdrawn.
  ,, Eastern Counties Railway opened to Colchester.
  .. Great Western Railway opened to Plymouth
1849 Shrewsbury and Birmingham Railway opened
1850. Chester and Holyhead Railway opened
1874. Last of the mail-coaches the Thurso and Wick Mail
      gives place to the Highland Railway.

INDEX

Coach-proprietors (*continued*) —

 Taylor Isaac, of Shrewsbury, ii 215, 216, 307

 Teather, Edward, of Carlisle ii 238

 Tubb, J., i 283-5

 Waddell, of Birmingham, ii 238

 Ward, Charles, ii 313-15

 Waterhouse, William, of the "Swan with Two Necks," Lad Lane, ii 196

 Webb Frederic, of Bolton, ii. 238

 Wetherald, J. & Co., of Manchester, ii 238 278

 Whitchurch, Best & Wilkins, of Brighton, i 312-15

 Willans, Wm, of the "Bull and Mouth," St Martin's-le-Grand, ii. 227

 Worcester, Marquis of (afterwards 7th Duke of Beaufort), ii 101

Coach travelling, on the roof, described by Moritz, 1782, i 99-102, by mail, 1798, described by Boulton, i 179, passengers booked in advance, i 321, miseries of early morning, i 325-32; about 1750, described in *Roderick Random*, i 333; courtesies to ladies, 1714, i 335, romance of, i 336, severe test of a gentleman, i 337, humours of coach-dinners, i 337-47, coach-breakfasts, i 347-51, social gulf between inside and outside passengers, i 351, described by De Quincey, i. 351-3, humour in, i 353, adventures described, i. 355, savage idea of humour, i 356 8, practical joking, i 357; outside the most desirable place in summer, ii 67, in 1772, ii. 48-65, in 1830, ii. 66-95, miseries of, in winter, ii 155-8, 169

"Comet' coaches, begin about 1811, i. 304-8

Commercial travellers, known successively as "riders," "bagmen," "travellers," "commercial gentlemen," "ambassadors of commerce," and "representatives," i. 56, come into existence about 1730, i 118; adventure of a, ii 328

"Common stage-waggons," a term specified by General Turnpike Act of 1766, i 204

Cornets-à-piston, popular with guards, i 280

Cresset, John, denounces stage-coaches, 1662, i. 26, 70-74

Darwin, Dr. Erasmus, prophesies railways and balloons, ii. 260, 282

"Derby Dilly," the, i 289

Dickens, Charles, on coach booking-offices, i. 322, on miseries of early morning travelling, i. 325-32, on coaching prints, ii 110 Christmas stories, ii 162, at the "Bull," Whitechapel, ii 231

Diligences, a species of Light Post-Coach, i 160, 287-92, originally fast, and carried three inside passengers only, i 287, became slow, i 288-90, Shillibeer's Brighton Diligence, i 290-92

"Double Horse," the, i 53

Eliot, George, foreshadows tube railways, ii. 282-5

Elizabeth, Queen, suffers from riding in carriage, i 5 ; prefers riding horseback, i 5

Fares, by stage-coach, a shilling for every five miles, 1684, i. 79 , London and Bath, £1 5s , 1667, i. 69 , Bath Flying Machine, 3d a mile, 1667, i 69 , London and Oxford, 12s , 1669, i 71 : 10s , 1671, i 71 , Liverpool Flying Machine, about 2½d a mile, 1758, i. 93 , reduced in competition on Brighton Road, 1762, i 284 , in competition with railways, 1838, ii 273 , Shrewsbury and London Long Coach, 18s , 1753, i 95 , Shrewsbury and London Caravan, 15s , 1750, i. 119 , Shrewsbury and London Stage, inside, £1 1s , 1753, i 119 , Shrewsbury and London Machine inside, 30s , 1764, i. 120 ; Newcastle and London, 1772, ii 63 , 1830, ii 67, 95 , reduced all round, 1834, ii. 187

Fares, Short stages, ii 189

 „ Waggon, from ½d. to 1d a mile, i 69, 139 , ¼d a mile, or 1s a day, i 120, 131

Floods, ii. 165-70

Fly Boats, i 140, ii. 130

 „ Vans, London and Falmouth, 1820, i 136-9

"Flying Coach," the first, 1669, i 69

"Flying Machines," the first, 1667, i 68 , described, i 68-93, 283-5

Flying Stage-waggon, London and Shrewsbury, in 5 days, 1750, i 118

Gamon, Sir Richard, legislates on coaching, i 206-8

Gay, John, the Poet, his *Journey to Exeter*, 1715, i 28-33

Goods, carriage of, by pack-horses, i. 106-111, ii 124 by sledges, called "Truckamucks," i 107 pack-horses partly replaced by waggons about 1730, i 117 , cost of carriage, 1750, i 135 , by road and canal, about 1830, i 140, carrying firms, ii 123-43, 207-10

Guards, generally, "shoulder" fares and "swallow" passengers, ii 200-203

Guards of mails, not to fire off blunderbusses unnecessarily, i 209 ; servants of General Post Office, i 249 , gross excesses of early, i 250-52 , Post Office responsible for excesses, i 251 ; how armed and equipped, i 251-60 , extravagant behaviour restricted, i 252 , appointments eagerly sought, i 252 , salary small, 10s 6d. weekly, i 253 , "tips" render appointments valuable, i 253 ; illegal purveyors of game, i 254 , trusted and confidential messengers, i 255 ; as smugglers, i 256 , bravery of, and devotion to duty, i 256 , number of, i 256 , responsibilities of, i 258 , purveyors of news, i 259 , their duties, i 261 , instructions to, i 262 , prosperity of, i 262 , position poor on cross-country mails,

Old-time travellers (*continued*).—

Denmark, Prince George of, visits Petworth, ii 332

De Quincey, Thomas, on contempt of inside passengers for outsides, i 210, 351-3 prefers outside of coaches, ii. 67

Dugdale, Sir William, mentions Birmingham coach of 1697, i 77

Fiennes, Celia, in Lancashire, 1691, surprised at finding sign-posts, i. 115

Gay, John (the poet) *A Journey to Exeter*, 1715, i. 28-33

Hawker, Col, on travelling in 1812, i 245, on cost of journey, London to Glasgow, 1812, ii. 1-3, 4, on "Bull and Mouth" inn, 1812, ii 227

Johnson, Dr, i 52-3

Macready, William C. (the actor), on incredibly slow journey, Liverpool to London, 1811, i 294

Moritz, Rev C H, on miseries of outside passengers, 1782, i 98-102

Murray, Rev James, describes a journey from Newcastle-on-Tyne to London 1772, ii 48-65

Parker, Edward, on miseries of coach journey from Preston, Lancashire, 1662, i. 25-63

Pepys, Samuel, often loses the road, i. 112

Somerset, Dukes of, and Petworth, ii. 332

Sopwith, Thomas, on discontinuance of York Mail, ii. 39

Sorbière, Samuel de, on waggoners, 1663, i 127

Swift, Jonathan, Dean, his couplets for inn signs on Penmaen-mawr, i 21, on horseback journey, Chester to London, 1710, i. 33, 73; on journey London to Holyhead and Dublin, 1726, i 33, diary of journey, London to Holyhead, 1727, i. 34-47, epigram at Willoughby, i 46, travels by stage-waggon, i 132, on travelling, ii. 330

Taylor, John (the "Water Poet"), travels to Southampton, 1648, i 58-60

Thoresby, Ralph, travels by York stage to London, 1683, i 27, 73, finds the Hull to York stage discontinued for winter season, 1678, i 74, going horseback, often misses his way, i 112, describes custom of treating lady passengers in coaches, 1714, i 335, on spring festivities, 1714, ii. 346

Wesley, John, generally travelled horseback, i 47, describes his adventures, i 47-52, finds unpleasant company in a coach, i 293

Omnibuses, displace "short stages," ii 193, "Wellington," Stratford and Westbourne Grove, ii 235, of Richmond Conveyance Co, ii 296

Outside passengers first heard of, and probable origin of carrying, i 95, miseries of, i 98-102: first provided with seats, i 181, treated with contempt by inside passengers, i 210, 351-3, ii 181

Pack-horses, i 106-9, 111, 118, partly replaced by waggons about 1730, i 117 · pack-horse trains, ii. 124

Palmer, John, Post Office reformer, account of, i 148-80 (Appendix, Vol. I., p 359), proposes a service of mail-coaches, i 155, plan for, matured 1782, i 156, establishes first mail-coach, 1784, i 158; proposes to extend system to France, i 163, appointed Comptroller-General 1786, i 164, contentions with Postmasters-General, i 165-72, his character, i 166; betrayed by Bonnor, i 168, dismissed, i. 172, giant to, i. 173, death of, i 174, ancestry of, Appendix, Vol I, p 359, descendants, 359

"Parliamentary Horse," the, i. 218

"Patent Safety" coaches, i 309-16; ii 109

Pepys, Samuel, sets up a carriage, 1668, i. 11, in travelling, often loses the road, i. 112

"Pickaxe' team, *i e.* three horses, ii 270

Pickford & Co, i. 139; ii. 123-43, 208

    ,,   Matthew, ii. 125-7

    ,   Thomas, ii 125-7

Poor people, how they travelled, i 113, 131-3, 139, find it cheaper to go by rail, i. 144

Postboys, *i e* mail-carriers, i 146, 152; went toll-free, ii. 5

Postes, Master of the, i 14

Post-horses, State monopoly of, i 11-23; monopoly abolished, 1780, i 23, mileage charges for, i 15, increased, i 18

Postmaster-General, office of created, 1657, i 18

Postmasters, *i e* keepers of post-horses, i 15-18, 147

"Post Office of England" created, 1657, i. 17, re-established, 1660, i 22

Post Office, General, i 14-19, 20, 22-4, 46-180; declines Hancock's offer to convey mails by steam-carriage, ii. 268

Railways ·—

    Mails first carried by, 1830, ii 16, authorised to convey mails, 1838, ii 16; run York coaches off road, 1840, ii 39, run waggons off, ii 138; threaten coaching, ii. 208, projected railways criticised, 1838, ii. 209, ruin the early steam-carriages, ii 268; ridiculed, 1837, ii. 268, cut up the coach routes, ii 270-74, bad service of trains, 1838, ii 274, insolence of officials, ii 274-7, public dissatisfaction with, 1838, ii. 274-7, tube railways foreshadowed by George Eliot, ii 282-5

    Grand Junction, ii 141, 274

    Highland, ii 40

    Liverpool and Manchester, ii 16, 96, 138

    London and Birmingham (now London and North-Western), ii 141, 208, 222-5, 273, 278

Railways (*continued*) —
   London and Manchester, ii 16, 96, 138
       ,,   ,, Southampton (now London and South-Western), ii.
      17, 36, 209, 290
   Metropolitan extended to Aylesbury 1892, ii 281
   North British, ii. 40
" Ride and Tie," custom of, i. 54
Rippon, Walter, carriage-maker to Queen Mary, i 4
Roads, bad state of, 1568, i 5, dreadful condition in North Wales in
   eighteenth century, i. 20-22, Exeter Road described in 1752 as
   "dreadful," i 91, first General Highway Act, 1555, i 106, mere
   tracks and unenclosed 1739, i 111, not safe for solitary travellers,
   i. 115, gradually improve from 1700, i 117, growth of heavy
   traffic cuts them up, i. 123, ignorance of road-surveyors, i 123,
   legislation to protect, 1760, i 123-6; 1622-29, 194-6, 1752, i. 199-
   202, General Turnpike Act, 1766, i 202-5, improve generally,
   ii 3, shocking state of, between Carlisle and Glasgow, 1812, ii. 4,
   wear and tear of, by mails, ii. 4-9, and early steam-carriages,
   ii 262· vulgarised by modern "improvements," ii 326. terrible
   state of, in Sussex, ii 332, picturesqueness of, threatened by
   coming changes, ii 347
Robberies from coaches, ii 144-50
" Rumble-tumble," i. 96, 97, 99, miseries of travelling in the, i. 101,
   139
Rutland, Earl of, sets up a carriage, 1555

Shillibeer, George, his "Brighton Diligence," i 290-92; his omni-
   buses, ii. 193
Short stages, the, ii 188-93
" Short Tommy," the, ii 175
" Shouldering," i e stealing, fares, ii. 200-203
Sign-posts obligatory, 1690, i. 112
Silver, Anthony, carriage-maker to Queen Mary, i 3
Smollett, Tobias, i 108, 110; on travelling in 1748, i. 115-17, 334
Snowstorms, i 261, 261-9, ii 137, 157, 159-62, 166-9
Stage-coaches—*see* "Coaches stage-coaches'
Stage-waggons, established about 1500, i 2 *see* "Waggons"
Steam-carriages, 1823-38, ii 217, 260-68
Sunday, a day of rest, i. 29, 90
   ,, Trading Acts, i 196-9. ii 118
" Swallowing," i e. stealing, fares, ii 200-203

Talbot, the old English hound, i. 109
" Tantivy," meaning of the word, ii 185
" Tantivy Trot," coaching song, ii 185
Telegraph coaches established, from about 1781, i 300-303

Printed by Hazell, Watson & Viney, Ld, London and Aylesbury